WHEN THINGS GET TOUGH, THERE ARE TWO THINGS THAT MAKE LIFE WORTH LIVING: MOZART, AND QUANTUM MECHANICS.
VICTOR WEISSKOPF

IT IS OFTEN STATED THAT OF ALL THE THEORIES PROPOSED IN THIS CENTURY, THE SILLIEST IS QUANTUM THEORY. IN FACT, SOME SAY THAT THE ONLY THING THAT QUANTUM THEORY HAS GOING FOR IT IS THAT IT IS UNQUESTIONABLY CORRECT.
MICHIO KAKU

THE TRUTH ALWAYS TURNS OUT TO BE SIMPLER THAN YOU THOUGHT.
RICHARD FEYNMAN

JAKOB SCHWICHTENBERG

NO-NONSENSE QUANTUM MECHANICS

NO-NONSENSE BOOKS

First printing, February 2019

Copyright © 2019 Jakob Schwichtenberg
With illustrations by Corina Wieber

All rights reserved. No part of this publication may be reproduced, stored in, or introduced into a retrieval system, or transmitted in any form or by any means (electronic, mechanical, photocopying, recording, or otherwise) without prior written permission.

UNIQUE ID: CD70DB24FF3EE32436DC5AFB203E3CDB7DB8656C9498C6755013743A7EDCE662
Each copy of No-Nonsense Quantum Mechanics has a unique ID which helps to prevent illegal sharing.

BOOK EDITION: 2.1

Dedicated to my parents

Preface

Quantum Mechanics is almost a century old. There are already hundreds of books about it. But unfortunately, Quantum Mechanics is still not an easy subject.

I remember when I learned Quantum Mechanics for the first time I was utterly lost. I struggled for months to understand even the easiest parts of it.

And in some sense, this is entirely normal. For example, there is the famous quote usually attributed to Niels Bohr: "If you are not completely confused by Quantum Mechanics, you do not understand it." Or more recently, Roger Penrose explained that his "general attitude" to Quantum Mechanics is that "it makes absolutely no sense!"

So, learning Quantum Mechanics is necessarily a hard and confusing process, right?

No! I no longer think this is true and this is what motivated me to write this book.

Just to be clear: Quantum Mechanics is not an easy subject. It takes some time to get used to it. You need to learn a completely new formalism. But at the same time this new formalism is not as difficult as most authors want you to believe and you don't need to spent months in a state of confusion. Also, be reassured that my goal is not to promote some crackpot alternative to Quantum Mechanics in order to remove all "quantum paradoxes" and "quantum mysteries". Instead, a large part of this book is about the standard topics that also most other textbooks talk about. However, my goal is to introduce them as gently as possible and, in some sense, I wrote the book I wished had existed when I started learning Quantum Mechanics.

Now, of course, the crucial question is: What exactly makes this book different?

▷ First of all, it wasn't written by a professor. So this book is by no means an authoritative reference. It's more like a casual conversation with a more experienced student who shares with you everything he wished he had known earlier. I'm convinced that someone who has just recently learned the topic can explain it much better than someone who learned it decades ago. Many textbooks are hard to understand, not because the subject is difficult, but because the author can't remember what it's like to be a beginner[1].

▷ Also, this book focuses solely on the fundamentals and contains no fluff. Most other books on Quantum Mechanics bury the essential concepts behind tedious calculations and complicated formulas.

▷ Another aspect that makes this book unique is that contains lots of idiosyncratic hand-drawn illustrations. Usually, textbooks include very few pictures since drawing them is either a lot of work or expensive. However, drawing figures is only a lot of work if you are a perfectionist. The images in this book are not as pretty as the pictures in a typical textbook since I firmly believe that *lots* of non-perfect illustrations are much better than a few perfect ones. The goal of this book, after all, is that you'll understand Quantum Mechanics and not that I win prizes for my pretty illustrations.

▷ Moreover, my only goal with this book was to write the most student-friendly Quantum Mechanics book and not, for example, to build my reputation. Too many books are unnec-

---

[1] This is known as the "Curse of Knowledge".

essarily complicated because if a book is hard to understand it makes the author appear smarter.[2] Concretely this means, for example, that nothing is assumed to be "obvious" or "easy to see". Moreover, calculations are done step-by-step and annotated to help you understand faster.

[2] To quote C. Lanczos: "Many of the scientific treatises of today are formulated in a half-mystical language, as though to impress the reader with the uncomfortable feeling that he is in the permanent presence of a superman."

With that said, the actual content and structure of the book is also somewhat non-standard:

In the first part we will talk about the essential features of Quantum Mechanics and how we can describe them mathematically. We start with a "Bird's Eye Overview" of Quantum Mechanics and then gradually refine our understanding.

Afterwards, we will discuss concrete examples and applications. We will stick to the most fundamental examples that show how Quantum Mechanics works in practice. The thing is that in most more advanced examples we don't learn anything new. Instead, the same lessons are buried behind more complicated calculations. But I will also comment on several important advanced examples that everyone should have heard about.

Next, we will talk about alternative formulations of Quantum Mechanics.[3] Most other textbooks focus solely on the wave function formulation and ignore the alternatives. However, similar to how it makes sense to know the Lagrangian and Hamiltonian formulations of Classical Mechanics, these alternative formulations can be extremely useful. This is especially true when we try to understand what Quantum Mechanics *really* means, which is the topic of the second to last chapter.

[3] Again: Be reassured that I will not talk about crackpot theories but only well-established alternative formulations like the path-integral formulation.

In the final chapter, we will discuss where you can learn more depending on what aspect of Quantum Mechanics interests you the most.

So, without any further ado, let's begin. I hope you enjoy reading this book as much as I have enjoyed writing it.

Karlsruhe, June 2018 *Jakob Schwichtenberg*

PS: I update the book regularly based on reader feedback. So if you find an error I would appreciate a short email to errors@jakobschwichtenberg.com.

PPS: You can discuss the content of the book with other readers, ask questions and find bonus material at:
www.nononsensebooks.com/qm/part1,
www.nononsensebooks.com/qm/part2,
www.nononsensebooks.com/qm/part3.

Acknowledgments

Many thanks are due to all readers of earlier versions of this book who suggested improvements, especially Patrik Iannotti, Corina Wieber, Salvador Ortega and John D. Nelson.

Before we dive in, we need to talk about two things. The first one is the following crucial question:

Why should you care about Quantum Mechanics?

The strongest argument in favor of Quantum Mechanics is that experiments tell us that it is correct. Physicists and philosophers can argue all day long about how beautiful some theory is, but in the end, the only thing that matters is if it agrees with what we observe in experiments. And Quantum Mechanics certainly does. Similarly to how you need Classical Mechanics to describe how a ball rolls down a ramp, we need Quantum Mechanics to describe, for example, the hydrogen atom[4].

[4] We will talk about lots of quantum systems in Part II.

In addition, Quantum Mechanics is the little brother of Quantum Field Theory, which is the best theory of the fundamental interactions that we have. And it is almost impossible to understand Quantum Field Theory without learning the basic concepts in the simpler context of Quantum Mechanics first.

To summarize[5]: Quantum Mechanics is an essential tool in the repertoire of any competent physicist. While it will not help you to describe the ball that rolls down a ramp any better, it will allow you to describe new physical systems which you can't describe with Classical Mechanics.

[5] That was quick. As promised I gave my best to remove all fluff. Most books discuss in the first few chapter how Quantum Mechanics came about historically. This story is super interesting. But this is not a history book, and there are already lots of great books on the history of Quantum Mechanics. See, for example,

J. E. Baggott. *The quantum story : a history in 40 moments*. Oxford University Press, Oxford England New York, 2011. ISBN 978-0199566846

The second thing we need to talk about is the meaning of a few special symbols which we will use in the following chapters.

Notation

▷ Three dots in front of an equation ∴ mean "**therefore**", i.e., that this line follows directly from the previous one:

$$\omega \hat{=} \frac{E}{\hbar}$$
$$\therefore \quad E = \hbar\omega .$$

This helps to make it clear that we are *not* dealing with a system of equations.

▷ Three horizontal lines ≡ indicate that we are dealing with a **definition**.

▷ The most important equations and results are highlighted like this:

$$\boxed{i\hbar \partial_t \ket{\Psi} = -\frac{\hbar^2 \partial_i^2}{2m} \ket{\Psi} + V(\hat{x}) \ket{\Psi}} \qquad (1)$$

▷ We often use the shorthand notation $\partial_i \equiv \frac{\partial}{\partial i}$ for the **partial derivative** where $i \in \{x, y, z\}$.

▷ It is conventional in Quantum Mechanics that we denote **operators** using a **hat**: $\hat{O}$. This notation helps to make clear when we are dealing with an operator and when with an ordinary number. However, often it is clear from the context that we are dealing exclusively with quantum operators and to unclutter the notation we neglect the hats.

▷ Moreover, square brackets denote the **commutator** of two operators $\hat{A}$ and $\hat{B}$: $[\hat{A}, \hat{B}] \equiv \hat{A}\hat{B} - \hat{B}\hat{A}$.

▷ ϵ_{ijk} denotes the three dimensional Levi-Civita symbol which is defined as follows

$$\epsilon_{ijk} = \begin{cases} 1 & \text{if } (i,j,k) = \{(1,2,3),(2,3,1),(3,1,2)\} \\ 0 & \text{if } i = j \text{ or } j = k \text{ or } k = i \\ -1 & \text{if } (i,j,k) = \{(1,3,2),(3,2,1),(2,1,3)\} \end{cases} \qquad (2)$$

That's it. We are ready to dive in. (After a short look at the table of contents).

Contents

Part I What Everybody Ought to Know About Quantum Mechanics

1 Birds-Eye View of Quantum Mechanics   25

2 Essential Quantum Features   33
    2.1 The Heart of Quantum Mechanics 33
    2.2 What does "Quantum" mean? 38
    2.3 Uncertainty 43

3 The Quantum Framework   47
    3.1 Intermezzo: Essential Statistics 51
    3.2 Statistical Tools in Quantum Mechanics 54
    3.3 Wave Functions 60
    3.4 Quantum Operators 66
        3.4.1 Intermezzo: Symmetries 67
        3.4.2 The Most Important Quantum Operators . . 71
    3.5 How Operators Influence Each Other 74
    3.6 How Quantum Systems Change 75
    3.7 Why Quantum Mechanics is About Waves 79
    3.8 Intermezzo: Eigenvectors and Eigenvalues 82
    3.9 Angular Momentum 83
    3.10 Spin 85
    3.11 Quantum Numbers 90

4 The Classical Limit   93

5 Summary   97

Part II Essential Quantum Systems and Tools

6 Tricks and Ideas We Need All the Time — 105
- 6.1 Let's Separate Time and Space 106
- 6.2 Why Quantum Waves are Smooth 108
- 6.3 Classification of Solutions 109

7 Quantum Mechanics in a Box — 113
- 7.1 The Infinite Box 113
- 7.2 The Finite Box 118
- 7.3 The Hydrogen Atom 124

8 Scattering off a Box — 127
- 8.1 The Step Potential 130
  - 8.1.1 $E < U$ 132
  - 8.1.2 $E > U$ 133
- 8.2 The Box Potential 134

9 Harmonic Quantum Mechanics — 137
- 9.1 The Magical Method 140

10 Quantum Systems with Spin — 147
- 10.1 Spin Measurements 149
- 10.2 Spin Addition 154

11 Further Systems — 157

12 When the Going Gets Tough, the Tough Lower Their Standards — 163
- 12.1 Perturbation Theory 164
  - 12.1.1 General Perturbation Formulas 165
  - 12.1.2 The Perturbed Infinite Box 170
- 12.2 What Other Tools Do We Have? 171

Part III What Your Professor is Not Telling You About Quantum Mechanics

13 Mathematical Arenas — 181

14 The World Beyond Wave Functions — 193
- 14.1 The Pilot Wave Formulation 193

	14.2	Path Integrals	198
		14.2.1 The Origin of the Classical Path	209
	14.3	Phase Space Quantum Mechanics	213
	14.4	Heisenberg Formulation	223
	14.5	Which Formulation Is The Best?	225

15 What does it all mean?   227

16 Get an Understanding of Quantum Mechanics You Can Be Proud Of   235

One Last Thing

Part V Appendices

A Taylor Expansion   245

B Fourier Transform   249

C Delta Distribution   253

Bibliography   257

Index   261

Part I
What Everybody Ought to Know About Quantum Mechanics

"The universe is full of magical things patiently waiting for our wits to grow sharper."

Eden Phillpotts

PS: You can discuss the content of Part I with other readers, find exercises to check your understanding and give feedback at www.nononsensebooks.com/qm/part1.

Now it's time to get serious. Our plan for the following chapters is the following:[6]

The first chapter in this part of the book is a bird's-eye view of Quantum Mechanics. The goal here is not to understand everything immediately but to get an overview as quickly as possible. Afterwards, we discuss all concepts in more detail. We start with a short discussion of the most famous quantum experiment which is known as the double slit experiment. This experiment tells us that we need waves to describe particles. From this result, we can conclude that for some fundamental quantities only a discrete set of values are allowed. We say they are quantized. In addition, we can understand that there is a fundamental uncertainty in quantum systems. For example, it is impossible to measure the momentum and the location of a particle at the same time with arbitrary precision.

Then we start to develop a framework that allows us to describe these features of quantum systems in mathematical terms. The first essential ingredients are a new type of object, called "kets" which we use to describe the system in question and operators that yield, for example, the momentum of the system if we act with them on the ket. The quantum uncertainty tells us that we need statistical tools and this observation leads us to another type of object, called "bras". However, bras are not really something new but simply "conjugated" kets. These new objects then also allow us to understand that the famous wave functions are merely a convenient way to write down the coefficients which we get if we expand a ket in the position basis.

Afterwards, using group theory and Noether's theorem we derive how the fundamental quantum operators explicitly look like.[7] The explicit form of the quantum operators allows us to derive the two most famous equations of Quantum Mechanics: the canonical commutation relation and the Schrödinger equation. The former encodes the quantum uncertainty while the latter describes how quantum systems evolve in time. Also, we can then understand that the values which we can measure for basic quantities like momentum or energy correspond to the

[6] Don't worry if most things don't make sense to you here at first glance. The only goal here is that you get a general idea of what we will do in this first part. We will discuss all the concepts mentioned here in detail in the following sections.

[7] Group theory is the part of mathematics that allows us to describe symmetries.

eigenvalues of the corresponding quantum operators. After a short general discussion of operators and their eigenvalues we move on to another crucial physical quantity: angular momentum. In Quantum Mechanics angular momentum consists of two parts. One part describes the usual orbital angular momentum that describes how two objects revolve around each other. The second part describes something new: spin, which is some kind of internal angular momentum. In the final chapter of this part, we discuss the connection between Classical Mechanics and Quantum Mechanics explicitly.

The following image is our travel guide. Whenever you feel lost, come back here. But for the moment it is sufficient to take a short look, and it is not necessary to study it in detail.

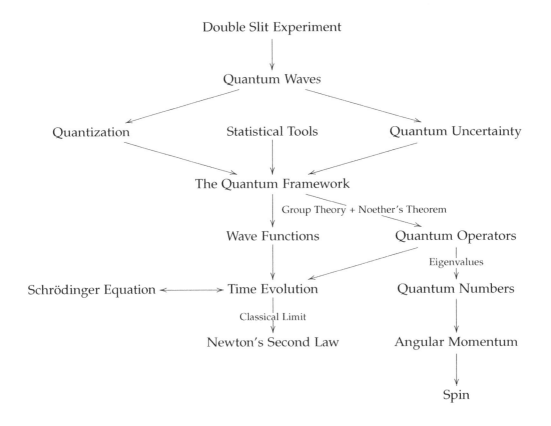

Now, as promised above, the following chapter is a whirlwind tour of Quantum Mechanics. We will talk about all fundamentally important concepts on just a few pages. It will be quite the ride. So please, get another cup of coffee, take a deep breath and then let's dive in.

1

Birds-Eye View of Quantum Mechanics

As mentioned already in the preface, Quantum Mechanics is at its heart, like most other theories in physics, quite simple. However, as always, certain applications can be extremely difficult and complicated. For this reason it's easy to loose the forest for the trees. To prevent this, in this chapter we will start by talking about all fundamental notions and concepts and by putting them into context. Afterwards, we will talk about the various concepts in more detail and gradually refine our understanding until we are ready for real-world applications.

There are three kinds of puzzle pieces that we need to master in order to get a proper understanding of Quantum Mechanics

1. Concepts (state vectors, operators, wave functions)
2. Equations (Schrödinger's equation, canonical commutation relations)
3. Tools (eigenfunctions, eigenvalues, expectation values)

There are, of course, lot's of additional concepts which are commonly discussed in the context of Quantum Mechanics. However, these are only important if we want to describe specific systems and not if we only want to understand the general structure of the theory.[1]

[1] We will discuss specific systems in Part 5.

Therefore, in this overview chapter, we focus solely on the fundamental concepts, tools and equations. Don't worry if not everything is immediately clear here. Our goal is only to get an overview and each idea mentioned in this chapter will be discussed later in more detail.

In Quantum Mechanics, we describe a given system using a **state vector** $|\Psi\rangle$. Such a state vector is commonly called a **ket**.

Quantities we can measure are represented in our quantum framework by **quantum operators** $\hat{O}$.[2] This fact is encoded in canonical commutation relations like

[2] Examples for quantities we can measure are: momentum $\hat{p}$, energy $\hat{H}$ or angular momentum $\hat{L}$.

$$[\hat{p}_i, \hat{x}_j] = \hat{p}_i \hat{x}_j - \hat{x}_j \hat{p}_i = -i\hbar \delta_{ij} \qquad (1.1)$$

Here $\hat{p}_i$ denotes the momentum operator in the i-direction, and $\hat{x}_j$ the position operator.[3] The most important point here is that the quantities like the momentum and the position are not represented by numbers as it is the case in Classical Mechanics. This is necessarily the case here since simple numbers always commute, e.g. $3 \times 5 - 5 \times 3 = 0$. Instead, the canonical commutation relation here tells us that in Quantum Mechanics quantities like momentum are represented by operators.

[3] For example, $\hat{p}_x$ yield the momentum in the x-direction and $\hat{x}_z$ yields the location on the z-axis.

The basic idea now is that we act with these measurement operators on the given state vector and, as a result, get the specific number that we can measure in an experiment.

$$\hat{O}|o_{\text{spec}}\rangle = o_{\text{spec}}|o_{\text{spec}}\rangle, \qquad (1.2)$$

where $\hat{O}$ is an operator and o_{spec} a number.

For example, let's imagine that we have prepared an electron such that it has momentum $3.21 \frac{\text{kg m}}{\text{s}}$. In our framework, we describe the electron using the ket $|3.21 \frac{\text{kg m}}{\text{s}}\rangle$. By acting with the momentum operator $\hat{p}$ on our state vector, we get the momentum of the system

$$\hat{p} |3.21 \frac{\text{kg m}}{\text{s}}\rangle = \left(3.21 \frac{\text{kg m}}{\text{s}}\right) |3.21 \frac{\text{kg m}}{\text{s}}\rangle . \quad (1.3)$$

However, we only get such a simple result if our system is in an **eigenstate** of the operator in question. For each operator there is a specific set of so-called **eigenstates**. These eigenstates are the basic building blocks of all state vectors that we use to describe quantum systems. Formulated more technically, we can use the eigenstates of a given operator as basis vectors for our state vectors.

An eigenstate is characterized by its property that if we measure the quantity associated with the operator in question, we always get the same specific result.[4] Mathematically, this means that we if we act with the operator on an eigenstate, we simply get a number times the original state vector and not something more complicated:

[4] This may seem trivial but in Quantum Mechanics it isn't.

$$\text{operator} \times \text{eigenstate} = \text{eigenvalue} \times \text{eigenstate} .$$

In contrast, a general state vector can be understood as a linear combination of eigenstates of the operator in question[5]

$$|\Psi\rangle = \sum_i a_i |o_i\rangle = a_1 |o_1\rangle + a_2 |o_2\rangle + \ldots , \quad (1.4)$$

[5] This is completely analogous to how we can write a general vector $\vec{v}$ in terms of basis vectors

$$\vec{v} = v_x \vec{e}_x + v_y \vec{e}_y + v_z \vec{e}_z.$$

where a_i are numbers, commonly known as **probability amplitudes**, which tell us with which **probability** we can expect to measure the values o_i. Concretely, the probability to measure the value o_2 is $|a_2|^2$. The general relationship between a probability amplitude and the corresponding probability is:

$$\text{probability to measure } o_i = |a_i|^2$$

.

What this means is that if we prepare our system completely equally multiple times and measure each time, say, the momentum, we will not always get the same result. Instead, we sometimes (with probability a_1) measure the value p_1 and sometimes other values like p_2.

To make sense of a system which is in a non-eigenstate, we need one additional tool known as **conjugated state vectors** or simply "**bras**" $\langle \Psi |$. A bra is a hermitian conjugated ket[6]

$$\langle \Psi | = |\Psi\rangle^\dagger = (|\Psi\rangle^\star)^T \tag{1.5}$$

[6] Hermitian conjugation means the combination of complex conjugation and transposition.

Using an eigenstate bra $\langle o_j |$, we can project out the required probability amplitude a_j from any general ket directly[7]

$$\langle o_j | \Psi \rangle = \sum_i a_i \langle o_j | o_i \rangle = \sum_i a_i \delta_{ij} = a_j. \tag{1.6}$$

[7] Here δ_{ij} denotes the Kronecker delta.

This works because eigenstates are orthogonal and therefore a product like $\langle o_j | o_i \rangle$ yields zero except for $i = j$.

Therefore, if we are interested in the probability to measure, say, the value $3.21 \frac{\text{kg m}}{\text{s}}$ for the momentum, we multiply our general ket with the corresponding eigenstate bra:

$$\langle 3.21 \frac{\text{kg m}}{\text{s}} | \Psi \rangle = \sum_i a_i \langle 3.21 \frac{\text{kg m}}{\text{s}} | o_i \rangle = a_{3.21}. \tag{1.7}$$

The probability is then $|a_{3.21}|^2$

Additionally, we can use bras to calculate **expectation values**. An expectation value tells us which value we will most likely measure for a given quantity. In our framework, we can calculate expectation values by sandwiching the corresponding operator between a bra and the corresponding ket which describes the system

$$\boxed{\text{expectation value} = \langle \Psi | \hat{O} | \Psi \rangle .} \tag{1.8}$$

Take note that here we don't have an eigenstate as a bra but instead the bra $\langle \Psi |$ corresponding to the ket $|\Psi\rangle$.

The main idea here is that by calculating such a product we get a sum over all possible values times the corresponding probabilities[8]

$$\langle\Psi|\hat{O}|\Psi\rangle = \sum_j \sum_i \langle o_j|a_j^\dagger \hat{O} a_i|o_i\rangle$$

$$= \sum_j \sum_i a_j^\dagger a_i \langle o_j|\hat{O}|o_i\rangle \quad (a_i \text{ and } a_j^\dagger \text{ are numbers })$$

$$= \sum_j \sum_i a_j^\dagger a_i \langle o_j|o_i|o_i\rangle \quad (\text{the operator } \hat{O} \text{ acts on } |o_i\rangle)$$

$$= \sum_j \sum_i o_i a_j^\dagger a_i \langle o_j|o_i\rangle \quad (o_i \text{ is a number})$$

$$= \sum_j \sum_i o_i a_j^\dagger a_i \delta_{ji} \quad (\langle o_j|o_i\rangle = \delta_{ji})$$

$$= \sum_i o_i |a_i|^2 .$$

[8] Using the expansion of our general ket in terms of eigenstates (Eq. 1.4) and the connection between bras and kets (Eq. 1.5), we can calculate directly the expansion of the corresponding bra:

$$\langle\Psi| = |\Psi\rangle^\dagger$$
$$= \left(\sum_i a_i |o_i\rangle\right)^\dagger$$
$$= \sum_i a_i^\dagger |o_i\rangle^\dagger$$
$$= \sum_i a_i^\dagger \langle o_i| .$$

In words this means we multiply each possible outcome o_i with the probability $|a_i|^2$ to measure this value and then sum over all these terms. The result of such a weighted sum helps us to understand which value we will most likely measure.

Another important observation is that for each operator, we get a different set of eigenstates.[9] This means, we can also expand a general state vector in terms of different eigenstates:

$$|\Psi\rangle = \sum_i b_i |\tilde{o}_i\rangle = b_1 |\tilde{o}_1\rangle + b_2 |\tilde{o}_2\rangle + \ldots, \quad (1.9)$$

where $|\tilde{o}_i\rangle$ are the eigenstates corresponding to a different operator $\hat{\tilde{O}}$.

[9] Take note that for operators which commute

$$\hat{A}\hat{B} - \hat{B}\hat{A} = 0$$

there is a common set of eigenstates.

The point here is that if we are interested, for example, in the momentum of the system, we expand our general state vector in terms of momentum eigenstates. And if we are interested in, for example, the energy of the system, we expand the state vector in terms of energy eigenstates. The numbers a_i, b_i we get by expanding a general state vector like this tell us directly the probabilities to measure a given result.

In general, there is not only a discrete set of possible outcomes but instead a continuous one. This means that we often have to replace our sum with an integral[10]

$$|\Psi\rangle = \int do\, a(o)\, |o\rangle \,. \tag{1.10}$$

[10] In some sense, an integral is simply a sum over a continuous set of values, here o.

Take note that our discrete set of coefficients a_i is now a function $a(o)$. However, the basic idea is still the same. For each possible measurement outcome o, we get a specific probability amplitude $a(o)$. The probability to measure the value o is $|a(o)|^2$.

The function $\psi(x)$ that we get by expanding a state vector in terms of position eigenstates

$$|\Psi\rangle = \int dx\, \psi(x)\, |x\rangle \tag{1.11}$$

is usually called the **wave function.**

Analogous to how we can describe a given vector $\vec{v}$ using the specific coefficients for some given basis, we can use $\psi(x)$ to describe the system.[11]

[11] In principle, a vector is little arrow sitting somewhere in space. Only by using a specific coordinate system and therefore specific basis vectors like $\vec{e}_x, \vec{e}_y, \vec{e}_z$, we can describe the vector using concrete numbers $\vec{v} = (3,2,3)^T$. Take note that for a different choice of coordinate system or a different set of a basis vectors (e.g., spherical basis vectors), we get different numbers describing the vector.

The time-evolution of quantum systems is described by the **Schrödinger equation**:

$$\boxed{i\hbar \partial_t |\Psi\rangle = -\frac{\hbar^2 \partial_i^2}{2m} |\Psi\rangle + V(\hat{x}) |\Psi\rangle \,.} \tag{1.12}$$

Formulated differently, the Schrödinger equation is the equation of motion for our state vectors and equivalently, for our wave functions

$$i\hbar \partial_t \psi(x) = -\frac{\hbar^2 \partial_i^2}{2m} \psi(x) + V(\hat{x}) \psi(x) \,. \tag{1.13}$$

With this equation at hand, the main task in Quantum Mechanics is to solve it for different systems, i.e. different $V(\hat{x})$. As a result, we get wave functions which describe the quantum system in question. The absolute square of such a wave function tells us where we can expect to find the particle in the system. For example:

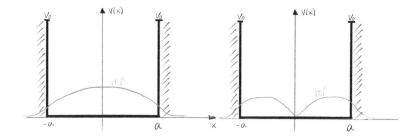

Moreover, we can use it to calculate which energy levels are possible and which momentum we can expect to measure.[12]

[12] We will discuss how exactly this works in Part 5.

After this certainly overwhelming first glance at what Quantum Mechanics is all about, let's take a step back and talk about the concepts introduced here in a bit more detail.

2

Essential Quantum Features

We will start with a short discussion of the three most important features of Quantum Mechanics:

1. **We need waves to describe particles,**
2. **quantization,**
3. **quantum uncertainty.**

The basic message to take away from this chapter is that nature on the fundamental scale works very different from what we expect from our everyday experiences. Afterwards, we will talk about how we can describe these unintuitive features using an intuitive physical framework.

2.1 The Heart of Quantum Mechanics

Richard Feynman once remarked in his lectures that one experiment *"has in it the heart of Quantum Mechanics. In reality, it contains the only mystery"* [1]. What magical experiment was he talking about?

[1] Richard Feynman. *The Feynman lectures on physics*. Basic Books, a member of the Perseus Books Group, New York, 2011. ISBN 978-0465025015

Well, astonishingly, it is extremely simple. All we need is a beam of electrons, a wall with two little slits in it and a screen behind the wall that detects the electrons. This experimental setup is known as the **double slit experiment**.

Before we discuss the experiment with actual electrons there are a few preliminary things we should talk about.

First of all, when we shoot a beam of, say, bullets towards a double slit and observe the pattern behind it, we will see something quite boring:

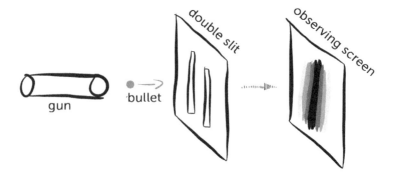

Most bullets end up in the middle of the screen and progressively fewer the more we move away from the middle. We can understand this pattern by considering the same experiment with one slit closed:

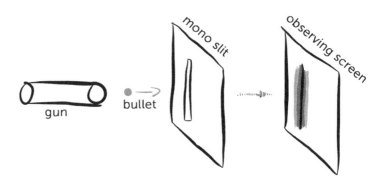

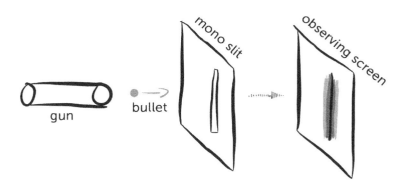

Then, if both slits are open, these two results simply add up.

In contrast, if we shoot waves, say water waves, towards a double slit we see something much more interesting

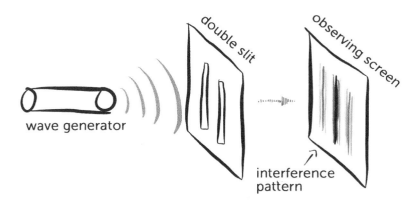

The pattern we see here on the screen is known as an **interference pattern**. This pattern arises because the wave goes through both slits and we can imagine that at each slit a new wave originates. As a result, behind the double slit, the two waves overlap and this leads to the interference pattern.

So, what does all this have to do with Quantum Mechanics?

Well, usually we think of particles like electrons as something like a bullet - only much tinier. Therefore, when we shoot electrons towards a double slit, we would expect a similar result as for bullets. The crazy thing now is that if we perform the double slit experiment with electrons, instead, we see an inference pattern on the screen!

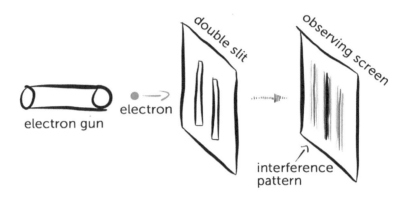

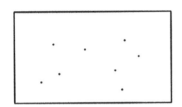

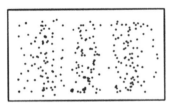

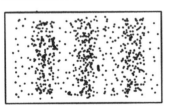

While for the waves we did measure the intensity (the height) of the waves, for the electrons we count how many of them end up at each particular point on the screen. We measure discrete events on the screen: "click, click, click,...". But if we plot how many electrons ended up at each location, we see the same pattern that we observe for water waves.

We can also observe the same behaviour for light. Since the concept of electromagnetic waves is somewhat familiar, this is maybe not so surprising. However, the important point is that if we shoot *tiny* amount of lights onto a double slit, we observe discrete chunks of light on the screen behind it. These light particles are called **photons**, and they are particles exactly in the same sense as electrons.

Physicists observed this behaviour experimentally and we still struggle to *understand* it. However, there is no problem in *describing* what is going on. And this is what Quantum Mechanics is primarily all about[2].

So the bottom line is:[3]

> **While electrons (and also all other elementary particles) are *discrete* things, we need *waves* to describe them.**

[2] However, we will also talk a bit about how we possibly can interpret this crazy behaviour in Chapter 15.

[3] This is not 100% correct. The thing is that waves are the simplest way to get a first grasp on what Quantum Mechanics is all about. We will discuss in Section 12.2 how we can describe quantum behaviour without waves.

Before we move one I want to warn you about one thing. In many books, there is lots of talk about things like the mysterious wave-particle duality. People wonder: How can an electron be a particle and a wave at the same time? Please ignore stuff like this. The situation is not as crazy as people want you to believe. Electrons, photons, and all other elementary particles are *particles*. Period. This is what every experiment tells us. Our detectors make "click, click, click". One convenient *mathematical tool* to describe the behaviour of these particles are waves[4]. This tool is convenient since everyone has already seen a water wave and can develop some intuition this way. However, these waves are only a mathematical tool. It is also possible to describe everything in Quantum Mechanics completely without waves[5]. So please, don't let yourself get confused by such discussions.

[4] The historical name for these waves is matter waves, which was introduced by the French physicist Louis de Broglie.

[5] We will talk about this in detail in Chapter 14.

In the following two sections we will talk about the two most important features of Quantum Mechanics. We can understand both directly when we use waves to describe particles.

2.2 What does "Quantum" mean?

In Classical Mechanics, we can measure for basic quantities like energy or angular momentum, in principle, any value we can imagine. A fundamental new thing in Quantum Mechanics is that this is no longer always true. Instead, in many quantum systems, our basic quantities arise only in discrete chunks. These discrete chunks are called **quanta**, and we say that basic quantities get **quantized**.

A famous example is electric charge. Any measurement of electric charge yields an integer multiple of the basic charge e:

$$Q = N \times e,$$

where N is an integer and e the charge of an electron[6]. Concretely, this means that a measurement of electric charge of some object in nature possibly yields $2 \times e$ or $977 \times e$, but never something like $2.12 \times e$ or $34.76 \times e$. In this sense, electric charge is quantized. Why this happens is a different question and, unfortunately, we don't know the answer yet[7]. But this is what makes Quantum Mechanics so interesting. There are still lots of fundamental things we do not understand.

No theoretical physicist predicted the quantization of electric charge. It was also not immediately discovered when we figured out electrodynamics. The thing is that the charges of macroscopic objects, say two metal spheres, are typically something like $Q_{sphere1} \approx 4.806$ C, $Q_{sphere2} \approx 5.01$ C, etc. These are huge numbers compared to the charge of a single electron e.

As a result, it seems as if we can produce spheres with arbitrary charges. However, this is an illusion. If we measure the electric charge of a macroscopic object very, very precisely, we notice that even such macroscopic charges are integer multiples of e:[8]

$$Q_{sphere1} \approx 4.806 \text{ C} \approx 3000 \cdot 10^{16} e$$
$$Q_{sphere2} \approx 5.01 \text{ C} \approx 3127 \cdot 10^{16} e. \qquad (2.1)$$

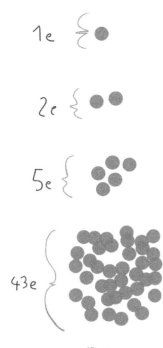

[6] $e = 1.602 \times 10^{-19}$ C

[7] One speculative possibility are special types of models known as grand unified theories. Theses models automatically explain the quantization of electrical charge through symmetry argument.

[8] Something like 10^{16} is a huge number:
$$10^{16} = 10 \times 10 \times \ldots 10$$
$$= 1000000000000000$$

The discrete chunks of electric charge are tiny compared to the charges of macroscopic objects. When we change the charge of a sphere what is happening is that we get a discrete sequence like

$Q_{sphere1} = 4.806529862399999555569404683720$ C

$Q_{sphere1+1e} = 4.806529862399999555585426449928$ C

$Q_{sphere1+2e} = 4.806529862399999556014482161367$ C

$Q_{sphere1+3e} = 4.806529862399999555633491748552$ C

$\vdots$

$Q_{sphere1+2948745599054545e} = 4.807002303525948763908673 48334$ C

If we are only concerned with the first four digits here, we can conclude that we can produce any value of the charge that we like. Only when we measure the charge *extremely* precisely, we notice that the change of the charge happens in discrete steps[9].

Now, the same thing also happens for other basic quantities. In many quantum systems, we can no longer measure any value for the energy or angular momentum[10]. Instead, we can again notice that they *only* appear in discrete chunks and are therefore quantized.

Famous examples are the energy levels of electrons in atoms. Classically, the electrons can have any arbitrary energy. Nothing in Classical Electrodynamics tells us otherwise. However, in Quantum Mechanics only a discrete set of energy levels is allowed. We can observe this by measuring the radiation which the atom emits whenever an electron jumps from one energy level to a lower one. The energy of the radiation is precisely the energy difference between the energy levels. There are no possible values in-between. The energy of the electron does not change continuously from one value to another. Instead, it jumps directly. This is where the notion of **quantum jumps** comes from.

Okay, what does all this have to do with the fact that we need waves to describe particles?

Well, waves naturally quantize quantities!

[9] We can understand this quantization since when we charge a sphere we add or remove electrons from it. Each additional electron contributions charge e. This is why we have discrete steps here. However, a mystery is why, for example, protons do not have some random value of the charge, but exactly the charge $-e$.

[10] In speculative theories like Loop Quantum Gravity, even space and time are quantized.

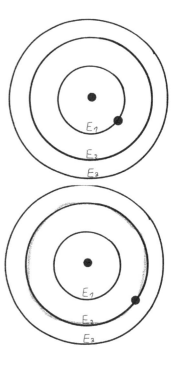

To understand this, let's consider a rope which we hold under constant tension using two hands:

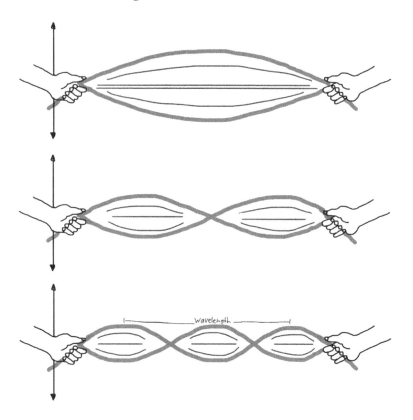

The thing is, no matter how the two hands try to make the rope vibrate, the rope will only vibrate with a quantized set of modes. The two hands fix the rope at both ends. As a result of this constraint, the rope can only vibrate with fixed modes. The modes between this fixed set of modes are physically impossible:

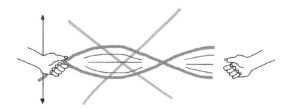

The same thing now also happens when we describe particles using waves. For example, let's consider an electron confined in a box[11]. If we describe the electron using a wave, we notice that only very particular wave shapes are allowed[12]:

[11] We will discuss this example in detail in Chapter 7.

[12] This is a result of the boundary conditions imposed by the box and is completely analogous to the rope example above.

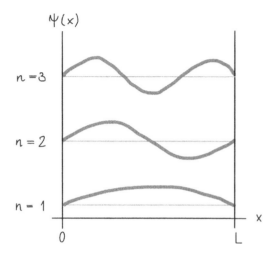

The **frequency** of the wave associated with a given particle is directly related to its energy[13]

$$\nu = \frac{E}{h}, \qquad (2.2)$$

where h is a constant known as the Planck constant, E the energy of the particle and ν the frequency of the corresponding wave[14]. For our particle in the box, this means that it can't have arbitrary energy values. Since only a discrete set of modes are allowed, it follows that only a discrete set of energy values is allowed. This is how energy becomes quantized in Quantum Mechanics!

[13] We will discuss the origin of this formula later in the book. For the moment we are only interested in the basic message and want to build some intuition.

[14] In slogan form: The higher the frequency, the higher the energy.

Similarly, we can understand why there are only discrete energy levels in atoms. As a naive picture imagine electrons orbiting the nucleus. Classically any orbit is allowed. However, as soon as we describe electrons with waves, this is no longer the case. The electron orbits now correspond to **standing waves** around the nucleus. For a fixed radius r we are effectively again dealing with something like a box[15]. The wave must coincide at

[15] Imagine in the picture of the box above that the points $x = 0$ and $x = L$ coincide. The fixed radius r is known as the **Bohr radius** and means physically the most probable distance between the nucleus and the electron.

the endpoint and the starting point and therefore, again, only specific wave modes are possible:

This is why the energy levels in atoms are quantized![16]

To summarize: Many familiar quantities like energy are not always continuous. In certain quantum systems, they only appear in discrete chunks[17].

Whenever this is true, we say the corresponding quantity is quantized. This fact is not observable in everyday life because the intervals between the allowed values are tiny. The quantization is a consequence of the "wave nature" of particles[18].

---

[16] This explanation is known as the **Bohr model**. Be warned that it is not entirely correct, outdated and works only good for the hydrogen atom. The details in a real atom are, of course, a lot messier. However, the Bohr model still gives a nice intuitive feeling of how quantization happens in Quantum Mechanics.

[17] Take note that, for example, the energy of a *free* particle is not quantized. The energy is only quantized for bounded systems like a box. However, the electric charge is always quantized. We will discuss this in a bit more detail in Chapter 5.

[18] As far as we know, this is *not* true for the quantization of electric charge, as far as we know. As already mentioned above, there is currently no generally accepted explanation for this particular quantization. However, still, it is a good example of what we mean by the word "quantization".

2.3 Uncertainty

The second crucial feature of Quantum Mechanics is that there is an *intrinsic* uncertainty we seemingly can't get rid of.

A somewhat naive but helpful explanation of the quantum uncertainty goes as follows:[19]

[19] The quantum uncertainty is often also called **Heisenberg uncertainty** and this is actually how Werner Heisenberg liked to explain it.

Imagine we want to measure the position of an electron by taking a picture of it. To take this picture we need to bounce light off the electron's surface since without light we can't see the electron. Such a picture reveals the position of the electron. However, unavoidably through the light, we push the electron a bit. As a result, the momentum of the electron gets changed each time we measure its position. What this means is that we can't know the position and momentum of an electron at the same time with arbitrary precision. We have to decide. Either we want to know the position exactly, but then are somewhat uncertain about its momentum. Alternatively, we want to measure its momentum and are then a bit uncertain about its location. There is a limit to how much we can know about an electron.

Again you may wonder: What does this have to do with our wave description of particles?

Good question. In the last section, we saw how quantization is a direct result of this new description. Now we will see that the quantum uncertainty is also a *direct* consequence of the "wave nature" of particles!

Once more, let's consider a rope. We can generate a wave in a long rope by shaking it rhythmically up and down:

Now, if someone asks us: "Where precisely is the wave?" we wouldn't have a good answer since the wave is spread out. In contrast, if we get asked: "What's the wavelength of the wave?" we could easily answer this question: "It's around 6cm".

We can also generate a different kind of wave in a rope by jerking it only once.

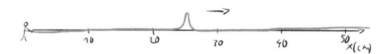

This way we get a narrow bump that travels down the line. Now, we could easily answer the question: "Where precisely is the wave?" but we would have a hard time answering the question "What's the wavelength of the wave?" since the wave isn't periodic and it is completely unclear how we can assign a wavelength to it.

Analogously, we can generate any kind of wave between these two edge cases. However, there is always a trade-off. The more precise the position of the wave is localized, the less precise its wavelength becomes and vice versa.

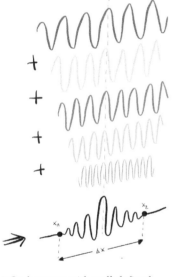

The thing is that we can think of a localized wave as a **superposition** of dozens of other waves with well-defined wavelengths[20].

If we add lots of waves with different wavelengths they will average out almost everywhere. Only within a small region, we can arrange the waves such that all of them add up don't cancel each other.

This is true for any wave phenomena. Since in Quantum Mechanics we describe particle using waves, it also applies here. In Quantum Mechanics, the **wavelength** is in a direct relationship

[20] Such waves with well-defined waves lengths are known as **plane waves**. The expansion of a general bump in terms of such plane waves is exactly the idea behind the Fourier transform. For more information on this see Appendix B. The uncertainty we end up with this way is a general feature of waves and known as **bandwidth theorem**.

to its momentum
$$\lambda = \frac{h}{p} \qquad (2.3)$$
The larger the momentum p, the smaller the wavelength λ of the wave that describes the particle. A spread in wavelength, therefore, corresponds to a spread in momentum.[21]

[21] Once more we use a formula here without any derivation. We do this to get some intuitive understanding of quantum waves first and postpone the actual derivation to a later chapter.

In physical terms this means exactly what we talked about above: We can't know the location and momentum of particles with arbitrary precision:

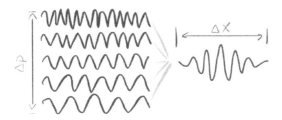

So, as a result, we end up with an uncertainty relation that tells us:

> **The more precisely we determine the location of a particle, the less precisely we are able to determine its momentum and vice versa.**

Before we move on, let's summarize what we learned in the previous sections.

The double-slit experiment tells us that we need *waves* to describe *particles*. The "wave nature" of particles then led us to two essential features of quantum systems: quantization and uncertainty.

Since within a box only special waveforms are allowed, we end up with a discrete set of allowed states. In physical terms, this means that, for example, within a box only a discrete set of energy values is allowed and we say that energy is quantized.

Moreover, we learned that for waves there is always a trade-off between the localization of the wave and how well we can determine its wavelength.

We can think of a localized wave bump can as a sum of many waves with well-defined wavelengths. In other words, a localized waveform corresponds to a superposition of waves with different wavelengths and therefore not to one specific wavelength. In turn, a waveform with well-defined wavelength (a plane wave) is not localized anywhere but spreads out all over space.

In Quantum Mechanics the wavelength is in a direct relationship to the momentum. In addition, the localization of the wave tells us, well, where our particle is.

Taken together, what all this tells us is that thanks to their "wave nature" we can't measure the location and the momentum of a particle at the same time with arbitrary precision.

Now it's time to think about how we can develop a framework that allows us to describe all these weird features of quantum systems.

3

The Quantum Framework

We will start this chapter by developing a general framework to describe nature. Afterwards, we will see how easily we can describe Quantum Mechanics using this general framework.

The most important ingredient that we need is *something* that describes our physical system in question. We invent a new mathematical symbol that does the job: $|\Psi\rangle$.

Everything about the physical system is encoded in $|\Psi\rangle$. The standard name for this type of object is **state vector** or **ket**[1].

[1] The name ket will make sense in a moment.

Okay, admittedly this sounds horribly abstract. But please don't run away. I promise this notation is clever and will make a lot of sense in a moment.

Each different state our system can be in corresponds to a different ket $|\Psi_1\rangle, |\Psi_2\rangle, \ldots$.

Think about an atom. An electron surrounding a nucleus can have different energies. It can be in the ground state (the state with the lowest energy), but it can also be in an excited stated.

In our framework, for example,

▷ $|\Psi_1\rangle$ denotes the configuration with the electron in the ground state and

▷ $|\Psi_2\rangle$ the configuration with the electron in the next highest excited state.

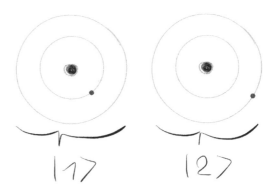

These are simply labels for the different states. For most systems, there are infinitely many possible states. For example, an electron at position $x = 0$ ($|x = 0\rangle$) is a different state than the same electron at position $x = 0.1$ ($|x = 0.1\rangle$). Since there are infinitely many possible locations, there are infinitely many states for a free electron. Again, this sounds a bit scary, but we will see in a moment how to deal with this situation.

So far so good. But now, how are these abstract objects related to anything we can measure in experiments?

Good question. There are only a few ingredients missing. The next thing that we need in our framework is a way to extract information from our kets, e.g. $|\Psi_1\rangle$. Say, we want to know the momentum of our system. To get this information from our abstract thing $|\Psi_1\rangle$, we "ask" it:

$$\hat{p}|\Psi_1\rangle = ? \tag{3.1}$$

The object $\hat{p}$ is called **momentum operator**. Again, we simply invented a mathematical symbol that does the job. The *operator*

$\hat{p}$ acts on our ket and as a result we get the momentum of our system:

$$\hat{p}|\Psi_1\rangle = p_1|\Psi_1\rangle. \tag{3.2}$$

Here the number p_1 is the momentum of the system in the configuration described by $|\Psi_1\rangle$. If we ask the same question when the system is in a different configuration, we get a different answer:

$$\hat{p}|\Psi_2\rangle = p_2|\Psi_2\rangle. \tag{3.3}$$

Yes, it's that simple! We act with $\hat{p}$ on the object that describes our system $|\Psi\rangle$ and get back the momentum that we can measure in an experiment.

Unfortunately, in Quantum Mechanics the situation is not really always that simple. In the previous section, we talked about one of the most important features of Quantum Mechanics: uncertainty. What this means for us here is that we will not always get such a simple answer as in the examples above. Often, a system that we try to describe has not simply the momentum p_1. Instead, there is always some variance in the answers. Even if we manage to prepare our system perfectly such that it has definitely momentum p_1, as soon as we ask questions about the location our measurements yield a range of values. Therefore, we need to add something to the framework that allows us to deal with uncertainty.

But what exactly does uncertainty mean here?

Well, in the example above it means that if we measure the momentum, we will not always get back the answer p_1. For 1000 equally prepared atoms we will sometimes measure p_1 and sometimes p_2. There is a variance in our measurements.

How can we implement this in our framework? Naively, we write

$$|\Psi\rangle = a|\Psi_1\rangle + b|\Psi_2\rangle, \tag{3.4}$$

where a and b are numbers that encode how often we measure p_1 and how often p_2, respectively. If we now act with $\hat{p}$ on this new ket, we get an answer we don't immediately understand

$$\begin{aligned}
\hat{p}\ket{\Psi} &= \hat{p}\Big(a\ket{\Psi_1} + b\ket{\Psi_2}\Big) &&\text{(using Eq. 3.4)} \\
&= a\hat{p}\ket{\Psi_1} + b\hat{p}\ket{\Psi_2} \\
&= ap_1\ket{\Psi_1} + bp_2\ket{\Psi_2}\,.
\end{aligned} \qquad (3.5)$$

The thing is that we can only understand a result like this in a *statistical* sense. This is what the quantum uncertainty demands from us. Results in Quantum Mechanics often only make sense in a statistical sense. Therefore, we need to add statistical tools to our framework.

Before we can do this, we first need to talk for a moment about statistical tools in general[2].

---

[2] If you are already familiar with the notions: **expectation value** and **standard deviation** feel free to skip the following section.

3.1 Intermezzo: Essential Statistics

One of the simplest but at the same time most important statistical tools is the so-called **expectation value**[3]. The expectation value of a quantity is the average value that we expect when we repeat a given experiment many times. We use it whenever we are forced to make probabilistic judgments[4].

[3] Alternative names for the expectation value are: expected value, mean value or simply the average. However, in the context of Quantum Mechanics, it is conventional to use the word "expectation value".

To understand it, imagine the following situation:

[4] Think: flipping a coin, or tossing a dice.

A friend offers to play a game. She flips a coin. If it lands on tails, she pays you $1.5. But if it lands on heads, you have to pay $1. By calculating the expectation value for this system, you can decide whether you should play this game.

▷ The probability that a coin lands on heads is $p_1 = 50\%$. In this case, the outcome for you is: $x_1 = -\$1$

▷ Equally, the probability that a coin lands on tails is $p_2 = 50\%$. In this case the outcome for you is: $x_2 = +\$1.5$

The expectation value is now defined as the sum over each outcome times the probability for the outcome

$$\text{expectation value} = \sum_i x_i P_i = x_1 P_1 + x_2 P_2$$
$$= -\$1 \times 50\% + \$1.5 \times 50\% = \$0.25. \quad (3.6)$$

So if we play this game many times, we will make a profit. On average we make $0.25 per game. For example, let's say you play the game only two times, win one time and lose the second time. For the win in the first game, you get $1.5. For the loss in the second game second game, you lose $1. In total, you have therefore won $0.5 in two games. This equals $0.25 per game.

Of course, here the system is so simple that we could have guessed this without calculating anything. But for more complicated systems the situations gets messy quickly. In quantum systems, there are lots of possible outcomes and, in general, we have different probabilities for different outcomes. In such cases, it is much harder to predict anything without such a useful notion.

To summarize: the expectation value is the expected outcome if we play the game many, many times. A common mathematical notation for the expectation value of a quantity x looks like this: $\langle x \rangle$.[5]

[5] Maybe you already noticed the similarity of this notation to our previous notation for the kets $|\Psi\rangle$. If not, don't worry, because the similarity will become a lot more obvious in a moment.

There is one additional statistical notion that we need all the time in Quantum Mechanics. It gives us information on how much the system *on average* deviates from the most probable outcome. In other words, this notion tells us how much our measurements are spread out. If it is zero, we measure the same value all the time. If it is non-zero, it's possible to measure different values.

The notion I'm talking about is called **standard deviation**. It tells us exactly how much our measurements (on average) spread around the most probable outcome (= the expectation value):[6]

$$\Delta x = \sqrt{\langle x^2 \rangle - \langle x \rangle^2}. \tag{3.7}$$

[6] This strange looking definition will make sense in a moment. And as an aside: The same definition without the square-root is known as **variance**.

The second term under the square root is simply the expectation value squared. For the first term, we calculate an expectation value again, but this time we square each possible outcome before we weigh it with the corresponding probability

$$\langle x^2 \rangle = \sum_i x_i^2 P_i. \tag{3.8}$$

Compare this with

$$\langle x \rangle^2 = \left(\sum_i x_i P_i \right)^2. \tag{3.9}$$

Yes, it makes a difference whether we square the whole sum or each term of the sum![7]

[7] For example, for $x = \{3, -5, 9\}$ we have

$$\sum_i x_i^2 = (3)^2 + (-5)^2 + (9)^2$$
$$= 115$$

whereas

$$\left(\sum_i x_i \right)^2 = (3 - 5 + 9)^2$$
$$= 49.$$

To understand this definition, let's consider the example from above again. The standard deviation here is

$$\Delta x = \sqrt{\langle x^2 \rangle - \langle x \rangle^2}$$

$$= \sqrt{\sum_i x_i^2 P_i - \left(\sum_i x_i P_i\right)^2} \qquad \text{(using Eq. 3.8 and Eq. 3.9)}$$

$$= \sqrt{\sum_i x_i^2 P_i - (\$0.25)^2} \qquad \text{(using Eq. 3.6)}$$

$$= \sqrt{((-\$1)^2 \times 0.5 + (\$1.5)^2 \times 0.5) - (\$0.25)^2}$$

$$= \$\sqrt{1.625 - 0.25} \approx \$1.17.$$

There are two important things to take note of:

▷ The result isn't zero. As I already emphasized above, it makes a difference whether we square each term in the sum or the complete sum. We can see this here explicitly. Otherwise, the standard deviation would always be zero.

▷ If we would only know that the expectation value of some game is $0.25 we are missing a lot of information. Naively, we could think since $0.25 is a tiny number that we are only betting tiny amounts of money. However, imagine that we make our example more extreme by proposing that if the coin lands on heads you have to pay $10000, and if it lands on tails you get $10000.5. The expectation value is again $\langle x \rangle = 0.25$. But the game is much more extreme now. I wouldn't like to play it. While the odds are still in your favor, there is also now a real chance that you lose *a lot* of money. This kind of information is encoded in the standard deviation. For this modified game the standard deviation is $\Delta x \approx \$10000.25$. We can now see immediately that a lot of money is at stake.

One more comment on the idea behind the definition of the standard deviation before we move on: The logic behind squaring each possible outcome in the first term on the right-hand side is to avoid that terms cancel because they have opposite signs. This is exactly what happens when we calculate the expectation value. In the coin example, if the outcome was heads we had to pay $1, which mathematically means $x_1 = -\$1$. Thus, in the formula for the expectation value, the two terms canceled almost completely since they have different signs. But now we want to get information about the spread in our measurements. Hence, we need to define a notion that exactly takes the *absolute distance* from the expectation value into account. We accomplish this by squaring each outcome and only then multiply each term calculated this way with the corresponding probability. We then also square the expectation value such that it has the same units as the thing we just calculated. Otherwise, we would compare apples with pears[8].

[8] In the example $\$^2$ with $.

Now, we have everything we need to get back to Quantum Mechanics.

3.2 Statistical Tools in Quantum Mechanics

After this short detour, let's recall what we already learned above:

We describe our system with an abstract object $|\Psi\rangle$ called a ket. If we want to know, for example, the momentum of the system, we act with the momentum operator $\hat{p}$ on this object. What we get this way is the value that we would measure in a momentum measurement. However, in general, we are dealing with uncertainty in quantum systems. Hence in general, we don't get back a simple number if we act with the momentum operator on the ket. For example, for our ket in Eq. (3.4) - which we recite here for convenience:

$$|\Psi\rangle = a |\Psi_1\rangle + b |\Psi_2\rangle , \qquad (3.10)$$

we have

$$\hat{p} |\Psi\rangle = a p_1 |\Psi_1\rangle + b p_2 |\Psi_2\rangle . \qquad (3.11)$$

In physical terms this means that we sometimes measure p_1 and sometimes p_2. The situation is therefore analogous to the coin example that we talked about in the previous section. Therefore, it makes sense to introduce the notion expectation value in

Quantum Mechanics. Before we can calculate expectation values we need to add a few more ingredients to our framework. The multitude of new concepts and ideas can be overwhelming at first. So don't worry if not everything is clear immediately. It takes some time to get used to new concepts. If you feel overwhelmed, just keep reading and everything will start to fall into place.

I will now simply list all new ideas and concepts that we need in the following. Afterwards, we will talk about them one after another in more detail. Ready?

1. The two ingredients which we need to calculate an expectation value is a list of possible outcomes plus the probabilities with which they occur. If we look at Eq. (3.11), we can see that for this particular ket the possible outcomes are p_1 and p_2. But what are the corresponding probabilities to measure these values? The correct answer is that for our example ket in Eq. (3.11) the probability to measure p_1 is $|a|^2$ and the probability to measure p_2 is $|b|^2$. So the correct formula for the momentum expectation value is

$$\langle p \rangle = |a|^2 p_1 + |b|^2 p_2, \qquad (3.12)$$

analogous to what we discussed in the last section.

2. In Quantum Mechanics, given a concrete ket $|\Psi\rangle$, we calculate the expectation value of an operator like $\hat{p}$ as follows:

$$\begin{aligned}\langle \Psi | \hat{p} | \Psi \rangle &= \left(a^* \langle \Psi_1 | + b^* \langle \Psi_2 | \right) \hat{p} \left(a | \Psi_1 \rangle + b | \Psi_2 \rangle \right) \quad \text{(using Eq. 3.10)}\\ &= \left(a^* \langle \Psi_1 | + b^* \langle \Psi_2 | \right) \left(a\hat{p} | \Psi_1 \rangle + b\hat{p} | \Psi_2 \rangle \right)\\ &= \left(a^* \langle \Psi_1 | + b^* \langle \Psi_2 | \right) \left(a p_1 | \Psi_1 \rangle + b p_2 | \Psi_2 \rangle \right)\\ &= |a|^2 p_1 \langle \Psi_1 | \Psi_1 \rangle + a^* b p_2 \langle \Psi_1 | \Psi_2 \rangle + b^* a p_1 \langle \Psi_2 | \Psi_1 \rangle\\ &\quad + |b|^2 p_2 \langle \Psi_2 | \Psi_2 \rangle\\ &= |a|^2 p_1 + |b|^2 p_2. \qquad (3.13)\end{aligned}$$

Here we used lots of new ideas that we will discuss in detail below.[9]

[9] Take note that we used that if we multiply a complex number z with its complex conjugate z^* we get the absolute value squared $|z|^2$. We will talk in a moment about the new symbols like $\langle \Psi_1 |$ that appear here and why the coefficients a and b appear complex conjugated here.

3. A new type object appears here in the definition of the expectation value in Eq. (3.13): $\langle \Psi |$. We call this kind of object a **bra**. This name together with the notion ket is a wordplay on the **bracket**s used here. A bra is nothing completely new. Given a ket, we can immediately compute the corresponding bra:

$$\langle \Psi | \equiv | \Psi \rangle^\dagger = (|\Psi\rangle^*)^T. \qquad (3.14)$$

The star $*$ here denotes complex conjugation and the superscript T transposition. The object $\dagger$ is called **dagger** and we use it to denote a transformation known as **hermitian conjugation**. We calculate the hermitian conjugate of an object by complex conjugating and transposing it. So in words, Eq (3.14) means that a bra is the hermitian conjugate of the corresponding ket.

4. In addition, we used in Eq (3.13) that our objects $\langle \Psi_1 |$ and $\langle \Psi_2 |$ are **normalized basis vectors**. This means they are orthogonal and thus, for example, $\langle \Psi_1 | \Psi_2 \rangle = 0$. Moreover, since they are normalized we have, for example, $\langle \Psi_1 | \Psi_1 \rangle = 1$. This is what we used to get to the last line in Eq (3.13).

So now let's talk about these ideas one after another in detail. We will start at the bottom of the list.

4. As already mentioned above, our kets like $|\Psi_1\rangle$ and $|\Psi_2\rangle$ that correspond to definite values of a given observable are **basis vectors**. This is in contrast to general kets which do not correspond to one definite value (here p_1 and p_2) of, say, the momentum operator but instead to a superposition of possible outcomes. So what we did above is to expand the general ket in terms of momentum basis kets[10]

$$|\Psi\rangle = a |\Psi_1\rangle + b |\Psi_2\rangle . \qquad (3.15)$$

This is completely analogous to how we can expand an arbitrary vector in terms of the basis vectors

$$\vec{e}_x = \begin{pmatrix} 1 \\ 0 \\ 0 \end{pmatrix}, \quad \vec{e}_y = \begin{pmatrix} 0 \\ 1 \\ 0 \end{pmatrix}, \quad \vec{e}_z = \begin{pmatrix} 0 \\ 0 \\ 1 \end{pmatrix}. \qquad (3.16)$$

[10] Momentum basis kets are those with definite momentum. A general state is a linear combination of momentum basis kets.

For example,

$$\vec{v} = \begin{pmatrix} 1 \\ 3 \\ 5 \end{pmatrix} = 1\vec{e}_x + 3\vec{e}_y + 5\vec{e}_z = 1\begin{pmatrix} 1 \\ 0 \\ 0 \end{pmatrix} + 3\begin{pmatrix} 0 \\ 1 \\ 0 \end{pmatrix} + 5\begin{pmatrix} 0 \\ 0 \\ 1 \end{pmatrix}. \tag{3.17}$$

A useful property of such basis vectors is that they are **orthogonal**[11]. Two vectors are orthogonal when their scalar product is zero. For example, $\vec{e}_x \cdot \vec{e}_y = 0$ and, in general, $\vec{e}_i \cdot \vec{e}_j = 0$ for $i \neq j$. In addition, they are **normalized**. This means that the scalar product of such a basis vector with itself yields one:[12] $\vec{e}_i \cdot \vec{e}_i = 1$.

This is why we used in Eq. (3.13) that $\langle \Psi_1 | \Psi_2 \rangle = 0$, $\langle \Psi_2 | \Psi_1 \rangle = 0$ and in addition, $\langle \Psi_1 | \Psi_1 \rangle = 1$, $\langle \Psi_2 | \Psi_2 \rangle = 1$.

[11] This is not always necessarily the case, but for the moment we assume that we only work with orthogonal basis vectors

[12] Again, this is not necessarily the case. But here we restrict ourselves to normalized basis vectors.

3. Next, we need to talk about the new kind of objects that we introduced above: $\langle \Psi |$. We introduce these bras specifically to be able to calculate expectation values and concrete probabilities. Both, expectation values and probabilities, are scalar quantities (i.e. ordinary numbers) and not vectors or matrices. So in other words, bras are tools that we add to our framework in order to combine them with out kets like $|\Psi\rangle$ and get back a scalar (i.e. an ordinary number). The whole point of bras is that if we combine them with a ket, we get back a number and not, for example, another ket or even an operator.

Again, we can understand this idea most easily by considering the analogous situation for the more familiar ordinary vectors. Here, the thing is that for each vector, we can define a "dual" vector such that the product of a vector and a dual vector yields a number. In other words, vector and dual vector together yield the scalar product. While a normal vector is a column vector, the corresponding dual vector is a row vector. Then, with the rules of matrix multiplication (row times column) a dual vector and a normal vector together really yield a number (and not another vector or a matrix):

$$\vec{v}^T \vec{v} = \begin{pmatrix} 1, 3, 5 \end{pmatrix} \begin{pmatrix} 1 \\ 3 \\ 5 \end{pmatrix} = 35.$$

A bra is completely analogous to a row vector while a ket is analogous to a column vector. The idea is that a bra and a ket together yield a number. In other words, this is a convenient way to write the **scalar product**[13]. The transformation that turns a column vector into a row vector is called transposition. Similarly, to get the corresponding bra from a given ket, we transpose it and *additionally* complex conjugate it. This conjugation is necessary since kets are, in general, complex[14]. For complex vectors, we need the additional complex conjugation because we want to interpret the scalar product of a vector with itself as the length of the vector squared. A complex result would make no sense. The complex conjugation makes sure we end up with something real. For this reason the scalar product for *complex* vectors is defined with an additional complex conjugation that makes sure we end up with a real number.[15]

This is also why we defined a bra in Eq. (3.14) as $\langle \Psi | \equiv |\Psi\rangle^\dagger = (|\Psi\rangle^*)^T$, and thus why a^* and b^* appear in Eq. (3.13).

2. Next, let's take another look at our first calculation of an expectation value in Quantum Mechanics (Eq. (3.13)), which I recite here for convenience:

$$\begin{aligned}
\langle \Psi | \hat{p} | \Psi \rangle &= (a^* \langle \Psi_1 | + b^* \langle \Psi_2 |)(a p_1 |\Psi_1\rangle + b p_2 |\Psi_2\rangle) \\
&= |a|^2 p_1 \underbrace{\langle \Psi_1 | \Psi_1 \rangle}_{=1} + a^* b p_2 \underbrace{\langle \Psi_1 | \Psi_2 \rangle}_{=0} + b^* a p_1 \underbrace{\langle \Psi_2 | \Psi_1 \rangle}_{=0} \\
&\quad + |b|^2 p_2 \underbrace{\langle \Psi_2 | \Psi_2 \rangle}_{=1} \\
&= |a|^2 p_1 \times 1 + a^* b p_2 \times 0 + b^* a p_1 \times 0 + |b|^2 p_2 \times 1 \\
&= |a|^2 p_1 + |b|^2 p_2.
\end{aligned} \quad (3.18)$$

The final result has exactly the structure that we talked about in the previous section. We weigh each possible outcome

---

[13] The notion scalar product means that we multiply two quantities and get back an ordinary number. "Scalar" is a different name for an ordinary number. There are other ways to multiply objects that do not yield an ordinary number like, for example, the cross product.

[14] We will see later why this is the case. For the moment we could say that we simply want to stay general and for this reason, include the complex conjugation which does no harm if the kets are real. But to spoil the surprise: Our kets are complex since kets that describe physical systems are solutions the Schrödinger equation. And solutions of the Schrödinger equation are, in general, complex.

[15] Because for $z = a + ib$ we have $z^* = a - ib$ and therefore
$z^* z = (a + ib)(a - ib) = a^2 + b^2$,
which is real.

(here p_1 and p_2) with the corresponding probability (here $P(p_1) = |a|^2$ and $P(p_2) = |b|^2$).

1. The last thing we need to talk about is why our probabilities are given by $P(p_1) = |a|^2$ and $P(p_2) = |b|^2$. Naively, we would have guessed that the probabilites to measure p_1 or p_2 are directly a and b, respectively. The thing is that kets which describe real physical situations are, in general, *complex*.[16] This means that coefficients like a and b are, in general, complex numbers. Since complex probabilities make no sense, we can't use coefficients like a and b directly. But instead, we can use $|a|^2$ and $|b|^2$, which are certainly real. So concretely, as already mentioned above, for our example ket in Eq. (3.11) the probability to measure p_1 is $|a|^2$ and the probability to measure p_2 is $|b|^2$. The coefficients like a and b that we get if we expand a ket in some specific basis are usually called **probability amplitudes**.

[16] We will see why this is the case in Section 3.6. As already mentioned above, the crucial point is that physical kets are solutions of the Schrödinger equation and solutions of the Schrödinger equation are, in general, complex.

With all this in mind, we can now tackle a second typical question in Quantum Mechanics: What's the probability of measuring, say, the value p_1. We already know that for our simple example from above it's $|a|^2$. But now we are interested in how we can extract this kind of information for any general system, no matter how complicated. The crucial idea is to use that the state of the system with definite momentum p_1 is $|\Psi_1\rangle$. Let's see what happens when we multiply our state $|\Psi\rangle$ with the bra that corresponds to $|\Psi_1\rangle$:

$$\langle\Psi_1|\Psi\rangle = \langle\Psi_1|\,(a\,|\Psi_1\rangle + b\,|\Psi_2\rangle)$$
$$= a\,\underbrace{\langle\Psi_1|\Psi_1\rangle}_{=1} + b\,\underbrace{\langle\Psi_1|\Psi_2\rangle}_{=0}$$
$$= a \times 1 + b \times 0 = a\,.$$

We get exactly the result that we wanted: a. To get the probability all we have to do is calculate the absolute square of this result.

This trick always works. A given ket $|\Psi\rangle$ possibly consists of dozens or even infinitely many term if we expand it in terms

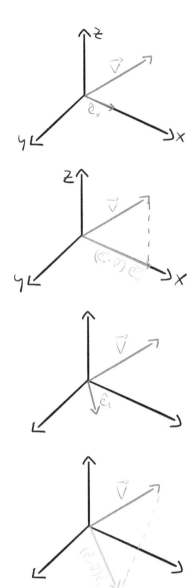

of basis states. But to get the probability to measure one specific momentum value, we simply have to multiply it with the corresponding conjugated basis state, e.g., $\langle \Psi_1 |$.

What we really do here is **project out** exactly the term that we want from the sum. Through the multiplication with $\langle \Psi_1 |$, all terms except for the one we are interested in yield zero since the basis vectors are orthogonal.

This whole procedure is completely analogous to how we can find out how much a given vector spreads out, say, in the z-direction. All we have to do is multiply the vector with the z basis vector $\vec{e}_z$:

$$\vec{e}_z \cdot \vec{v} = \vec{e}_z^T \vec{v} = \begin{pmatrix} 0 & 0 & 1 \end{pmatrix} \begin{pmatrix} 1 \\ 3 \\ 5 \end{pmatrix} = 5.$$

We can make the analogy even more explicit by using the notation from Eq. (3.17):

$$\begin{aligned} \vec{e}_z \cdot \vec{v} &= \vec{e}_z \cdot (1\vec{e}_x + 3\vec{e}_y + 5\vec{e}_z) \\ &= 1\vec{e}_z \cdot \vec{e}_x + 3\vec{e}_z \cdot \vec{e}_y + 5\vec{e}_z \cdot \vec{e}_z \\ &= 1 \times 0 + 3 \times 0 + 5 \times 1 = 5. \end{aligned} \quad (3.19)$$

3.3 Wave Functions

A question you certainly have at this point is: What does all this have to with the waves we talked about previously? Well, the **wave functions** everyone is talking about are simply the probability amplitudes for one specific basis. Again, let's consider the arbitrary ket $|\psi\rangle$ that describes our system. But now, let's say we are interested in possible locations and not the momenta. This means, we have to switch our basis and express $|\psi\rangle$ in terms of position basis states.

How can we do this?

The magical tools that help us to switch from one basis to another are called **projection operators**. Again, these projection operators aren't something completely new. Projection operators also exist for ordinary vectors. In the previous section, we used the most common basis vectors $\vec{e}_x, \vec{e}_y, \vec{e}_z$ (Eq. (3.16)). However, an equally good (orthogonal and normalized) choice for the basis vectors is[17]

$$\vec{\tilde{e}}_1 = \frac{1}{\sqrt{2}} \begin{pmatrix} 1 \\ 1 \\ 0 \end{pmatrix}, \quad \vec{\tilde{e}}_2 = \frac{1}{\sqrt{2}} \begin{pmatrix} 1 \\ -1 \\ 0 \end{pmatrix}, \quad \vec{\tilde{e}}_3 = \begin{pmatrix} 0 \\ 0 \\ 1 \end{pmatrix}. \quad (3.20)$$

[17] The factors $\frac{1}{\sqrt{2}}$ are normalization constants that make sure that our vectors have length 1.

How can we calculate how our vector $\vec{v} = (1,3,5)^T$ looks like in this basis?

We already discussed how we can find out how much a given vector spreads out in any given direction. For example, to find out how much $\vec{v}$ spreads out in the z-direction we multiplied it with $\vec{e}_z$ (Eq. (3.19)). Now Eq. (3.20) defines new axes relative to the old ones. Therefore, to find out how much $\vec{v}$ spreads out in the direction defined by, say, $\vec{\tilde{e}}_1$, we calculate the scalar product of the two vectors

$$\vec{\tilde{e}}_1 \cdot \vec{v} = \frac{1}{\sqrt{2}} (1,1,0) \begin{pmatrix} 1 \\ 3 \\ 5 \end{pmatrix} = \frac{4}{\sqrt{2}}.$$

Analogously, we can calculate how much $\vec{v}$ spreads out in the other two new directions

$$\vec{\tilde{e}}_2 \cdot \vec{v} = \frac{1}{\sqrt{2}} \begin{pmatrix} 1 \\ -1 \\ 0 \end{pmatrix}^T \begin{pmatrix} 1 \\ 3 \\ 5 \end{pmatrix} = \frac{-2}{\sqrt{2}}.$$

$$\vec{\tilde{e}}_3 \cdot \vec{v} = \begin{pmatrix} 0 \\ 0 \\ 1 \end{pmatrix}^T \begin{pmatrix} 1 \\ 3 \\ 5 \end{pmatrix} = 5.$$

This tells us that our vector $\vec{v}$ reads in the new basis (Eq. (3.20))

$$\vec{v} = \frac{4}{\sqrt{2}}\vec{e}_1 - \frac{2}{\sqrt{2}}\vec{e}_2 + 5\vec{e}_3 \triangleq \begin{pmatrix} \frac{4}{\sqrt{2}} \\ \frac{-2}{\sqrt{2}} \\ 5 \end{pmatrix}_{\text{new basis}}$$

The general method to rewrite a vector in a new basis is therefore

1. Calculate the scalar product of the vector with each new basis vector.
2. Multiply each result with the corresponding basis vector.
3. The vector in the new basis is the sum of all terms calculated in step 2.

So mathematically, we have

$$\vec{v}_{\text{new basis}} = \sum_i (\vec{e}_i) \underbrace{(\vec{e}_i \cdot \vec{v})}_{\text{a number}} . \qquad (3.21)$$

To convince you that this formula is really correct, let's again consider our example from above:

$$\vec{v}_{\text{new basis}} = \sum_i (\vec{e}_i)(\vec{e}_i \cdot \vec{v})$$
$$= (\vec{e}_1)(\vec{e}_1 \cdot \vec{v}) + (\vec{e}_2)(\vec{e}_2 \cdot \vec{v}) + (\vec{e}_3)(\vec{e}_3 \cdot \vec{v})$$
$$= \vec{e}_1 \frac{4}{\sqrt{2}} + \vec{e}_2 \frac{-2}{\sqrt{2}} + 5\vec{e}_3 \quad \checkmark$$

We use *exactly* the same method in Quantum Mechanics. We have the ket in the momentum basis

$$|\Psi\rangle = a|\Psi_1\rangle + b|\Psi_2\rangle . \qquad (3.22)$$

and want to calculate how it looks like in the position basis:

$$|\Psi\rangle = c|x_1\rangle + d|x_2\rangle . \qquad (3.23)$$

In other words, we want to calculate the coefficients c and d. For simplicity we assume here that only two locations x_1 and x_2 are possible. In addition, we use a more suggestive notation: $|x_1\rangle$ is the configuration of the system where we will definitely find our particle at location x_1, similar to how $|\Psi_1\rangle$ corresponds to the configuration with momentum p_1.

Using the algorithm we just discussed, we calculate

$$|\Psi\rangle = \sum_i |x_i\rangle \langle x_i|\Psi\rangle$$
$$= |x_1\rangle \langle x_1|\Psi\rangle + |x_2\rangle \langle x_2|\Psi\rangle$$
$$\equiv c|x_1\rangle + d|x_2\rangle,$$

where the **probability amplitudes in the position basis** are

$$c = \langle x_1|\Psi\rangle$$
$$d = \langle x_2|\Psi\rangle.$$

In general, there is a continuum of possible locations and not just a discrete set. Luckily, all we have to change is to replace the sum with an integral[18]:

$$|\Psi\rangle = \int dx\, |x\rangle \langle x|\Psi\rangle$$
$$\equiv \int dx\, \Psi(x)|x\rangle, \qquad (3.24)$$

where $\Psi(x) \equiv \langle x|\Psi\rangle$. This function $\Psi(x)$ is analogous to the coefficients we discussed previously (i.e. a, b, c, and d). But we now have one coefficient for *each* location x.[19]

The physical interpretation of $\Psi(x)$ is again that it's a probability amplitude[20]. This probability amplitude $\Psi(x)$ has a special name: **wave function**. Completely analogous to what we discussed before, if we take the absolute square of $\Psi(x)$ we get the probability to find the particle at a given location.[21]

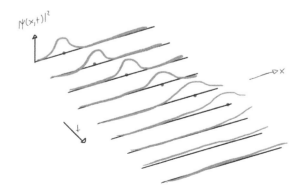

[18] An integral is, in some sense, the continuum limit of a sum. If we make the steps in a sum smaller and smaller, we end up with an integral.

[19] Take note that $|\Psi\rangle$ and therefore also $\Psi(x)$ both also depend on the time t. Here we suppress this dependence to unclutter the notation and since this time-dependence only become important later. But keep in mind that, in general, wave functions not only depend on the location but also on the time $\Psi(x,t)$.

[20] To make the connection to our previous notation explicit

$$\Psi(x_1) \hat{=} c$$
$$\Psi(x_2) \hat{=} d.$$

[21] This is not entirely correct. What we get is the probability to find the particle within the interval $[x, x+dx]$ since $\Psi(x)$ is defined under an integral. Thus to get a probability, we not only have to take the absolute square but also integrate over some region. In the image here the absolute square of a wave function at different points in time is shown. The absolute square indicate the probability to find the particle in question at different locations. The higher the amplitude of the absolute square of the wave function, the higher the probability. When we are 100% certain where our particle is, the wave function of the particle is the delta distribution (see Appendix C). In some sense, the delta distribution can be thought of as an infinitely thin/localized but also infinitely high function that, however, yields exactly one if we integrate it all over space.

The name *wave* function will make sense in Section 3.7 where we will see that functions like $\Psi(x)$ really behave like waves. One crucial observation is that everything we need to know about the system is encoded in the function $\Psi(x)$. This is analogous to how we usually only write down the coefficients of a vector $\vec{v} = (1,3,5)^T$ without referring to the basis explicitly: $1\vec{e}_x + 3\vec{e}_y + 5\vec{e}_z$. Similarly, we use in Quantum Mechanics $\Psi(x)$ and don't write $\int dx \Psi(x) |x\rangle$ all the time.

The trick we used in Eq. (3.24) is a very general one and works for any complete basis, not just the position basis. So instead of expanding our ket $|\Psi\rangle$ in terms of the position basis vectors $|x\rangle$, we can also expand it, for example, in terms of momentum basis states:

$$|\Psi\rangle = \int dp\, |p\rangle \langle p|\Psi\rangle$$
$$\equiv \int dp\, \Psi(p) |p\rangle \,. \quad (3.25)$$

The function $\Psi(p)$ is known as the **momentum representation** of the wave function.

Take note that by comparing the left-hand side with the right-hand in the first row of Eq. (3.24), we can conclude that the operator $\int dx\, |x\rangle \langle x|$ is the identity operator, i.e., an operator that does not change anything[22].

Analogously, from the first row in Eq. (3.25), we learn that the operator $\int dp\, |p\rangle \langle p|$ is also an identity operator $\int dp\, |p\rangle \langle p| \equiv 1$. This is important since it means that we can always insert $\int dx\, |x\rangle \langle x|$ or $\int dp\, |p\rangle \langle p|$ in our equations without changing anything. In a moment, we will talk about an example where this is indeed useful.

An important idea is now that using a given wave function $\Psi(x)$ we can immediately calculate important quantities like the expectation value[23]

---

[22] A general rule that is useful to remember is that bra times ket yields a number (or function): $\langle x|\Psi\rangle = \Psi(x)$. And ket times bra yield an operator: $|x\rangle \langle x|$. This is analogous to how the usual scalar product of two vectors yields a number: $\vec{v} \cdot \vec{w} \equiv \vec{v}^T \vec{w} \in \mathbb{R}$, while $\vec{v}\vec{w}^T$ yields a matrix. (Remember the general rule for multiplying matrices "row times column".)

[23] Using that $\langle \Psi| \equiv |\Psi\rangle^\dagger$ (Eq. (3.14)) and
$$|\Psi\rangle \equiv \int dx \Psi(x) |x\rangle$$
(Eq. (3.24)) we find immediately
$$\langle \Psi| \equiv \int dx \langle x| \Psi^\dagger(x) \quad (3.26)$$
Take note that we could equally write $\Psi^*(x)$ here instead of $\Psi^\dagger(x)$ since $\Psi(x)$ is an ordinary function and therefore transposing it makes not difference. Moreover, the symbol $\delta(x - x')$ that appears here is called delta distribution and is for integrals what the Kronecker delta δ_{ij} is for sums. For more information see Appendix C.

$$
\begin{aligned}
\langle\Psi|\hat{O}|\Psi\rangle &= \langle\Psi|\hat{O}\int dx\,\Psi(x)\,|x\rangle && \text{(using Eq. 3.24)} \\
&= \int dx'\,\langle x'|\,\Psi^\dagger(x')\hat{O}\int dx\,\Psi(x)\,|x\rangle && \text{(using Eq. 3.26)} \\
&= \int dx'\int dx\,\langle x'|\,\Psi^\dagger(x')\hat{O}\Psi(x)\,|x\rangle \\
&= \int dx'\int dx\,\Psi^\dagger(x')\hat{O}\Psi(x)\,\underbrace{\langle x'|x\rangle}_{=\delta(x-x')} \\
&= \int dx\,\Psi^\dagger(x)\hat{O}\Psi(x)\,. && (3.27)
\end{aligned}
$$

In the second to last step we used that if we act with $\hat{O}$ on our wave function $\Psi(x)$ we get back a number and we can move numbers around freely. In addition, the basic states $|x\rangle$ are orthogonal and normalized. Therefore, a scalar product like $\langle x'|x\rangle$ yields zero except when $x' = x$. This is analogous to how the scalar product between two basis vectors $\vec{e}_i \cdot \vec{e}_j$ yields zero, except when $i = j$. So when we integrate over x', all terms vanish except for the one with $x = x'$. This is what the Delta distribution $\delta(x - x')$ encodes, similar to how the Kronecker delta δ_{ij} picks exactly the element from a given sum where $i = j$.

We can now also perform the same steps as in Eq. (3.27) but without the operator $\hat{O}$ and with a bra that is different from the ket:

$$
\begin{aligned}
\langle\Phi|\Psi\rangle &= \int dx'\,\langle x'|\,\Phi(x')\int dx\,\Psi(x)\,|x\rangle && \text{(using Eq. 3.24 and Eq. 3.26)} \\
&= \int dx'\int dx\,\langle x'|\,\Phi^\dagger(x')\Psi(x)\,|x\rangle \\
&= \int dx'\int dx\,\Phi^\dagger(x')\Psi(x)\,\underbrace{\langle x'|x\rangle}_{=\delta(x-x')} \\
&= \int dx\,\Phi^\dagger(x)\Psi(x)\,. && (3.28)
\end{aligned}
$$

This is how the **scalar product** of two kets looks like if we use an explicit basis. Take note that this is completely analogous to the scalar product of two complex vectors in an explicit basis:

$$\vec{w}^* \cdot \vec{z} = \sum_i \sum_j (w_i^* \vec{e}_i) \cdot (z_j \vec{e}_j)$$
$$= \sum_i \sum_j w_i^* z_j \underbrace{\vec{e}_i \cdot \vec{e}_j}_{=\delta_{ij}}$$
$$= \sum_i w_i^* z_i \,.$$

Don't worry if not everything is perfectly clear at this point. Everything we discussed here will make a lot more sense as soon as we discuss explicit examples. The goal here is to get an overview and not to understand everything immediately.

Now, the two crucial questions that are still open are:

▷ How do these quantum operators we talked about all the time *really* look like?

▷ And secondly, how do quantum system evolve as time passes on?

This is what we talk about next.

3.4 Quantum Operators

So far, we talked about quantum operators in quite abstract terms. We simply introduced a symbol $\hat{p}$ and proposed that it does the job:

$$\hat{p}\,|\Psi_1\rangle = p_1\,|\Psi_1\rangle\,. \tag{3.29}$$

Luckily, there is one crucial idea that helps us go far beyond that. The idea I'm talking about is Noether's theorem[24].

Discussing the theorem in full detail would lead us too far astray. All we need is the basic message:

> **Symmetries lead to conserved quantities.**

[24] There are, in fact, two famous theorems by Emmy Noether. The one I'm talking about here is her first one that deals with *global* symmetries. Her second one is about *local* symmetries.

The most important examples are

▷ If the system does not change under rotations, we know immediately that angular momentum is conserved. In other words this means that if we can rotate our system without changing anything, angular momentum is conserved.

▷ If the system does not change under spatial translations $x \to x + \epsilon$, we know immediately that momentum is conserved. This means that if we can change the position of the whole system and nothing changes, momentum is conserved.

▷ If the system does not change under temporal translations $t \to t + \epsilon$, we know immediately that energy is conserved. Formulated differently, if the system behaved yesterday exactly as it does today, energy is conserved.

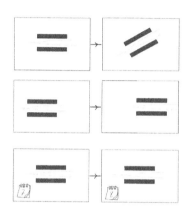

This is one puzzle piece. The second piece that we need is how symmetries are described mathematically.

You are probably wondering what all this has to do with Quantum Mechanics. As we will see after the next section the answer is: a lot! The mathematical ideas in next section are exactly what we need to find the explicit form of the quantum operators. Most importantly this explicit form allows us to derive the famous canonical commutation relation and the Schrödinger equation.

3.4.1 Intermezzo: Symmetries

The role and description of symmetries in physics is a huge topic[25]. So there is no way we can cover all the details here. However, I will try to emphasize the basic ideas that are necessary to understand Quantum Mechanics.

[25] In fact, I've written a whole book on exactly this topic:

Jakob Schwichtenberg. *Physics from Symmetry*. Springer, Cham, Switzerland, 2018. ISBN 978-3319666303

First of all: what is a symmetry?

Imagine a friend stands in front of you and holds an object in her hands. Then you close your eyes and she performs a trans-

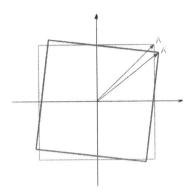

formation of the object (e.g. a rotation). Then you open your eyes again. If you can't tell if your friend changed anything at all, the transformation she performed is a symmetry of the object. For example, if she holds a perfectly round, single-colored ball in her hands, any rotation is a symmetry of the ball. In contrast, if she holds a box in her hand only very specific rotations are symmetries of the box. Doing nothing is always a symmetry.

The bottom line is:

> **A symmetry is a transformation that leaves the object in question unchanged.**

The part of mathematics which deals with symmetries is called **group theory**. A group is a set of transformations which fulfill special rules plus an operation that tells us how to combine the transformations. The rules are known as group axioms and we can motivate them by investigating an intuitive symmetry like rotational symmetry. We will not discuss details like this here since we don't need them for what follows.

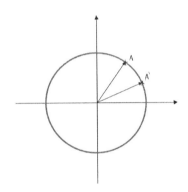

Also, we only need one special part of group theory, namely the part that deals with *continuous* symmetries. An example of a continuous symmetry is the one I just mentioned: rotations. Rotations are continuous because we can label them with a continuous parameter: the angle of rotation. In contrast, there are also *discrete* symmetries. The most famous examples are mirror symmetries.

There is one property that makes continuous symmetries especially nice to deal with: they have elements which are arbitrarily close to the identity transformation[26]. An arbitrary group has, in general, no element close to the identity.

[26] The identity transformation is the transformation that changes nothing at all. For example, for rotations, the identity transformation is a rotation by 0°.

Take, for example, the symmetries of a square. The set of transformations that leaves the square invariant are four rotations: a rotation by 0°, by 90°, by 180° and one by 270°, plus some mirror symmetries. A rotation by 0,000001°, which is very close to the identity transformation (a rotation by 0°), is not a symmetry.

Next, take a look at the symmetries of a circle. Certainly, a rotation by 0,000001° is a symmetry of the circle. Mathematically, we write an element g close to the identity I like this:

$$g(\epsilon) = I + \epsilon G \qquad (3.30)$$

where ϵ is, as always in mathematics, some really, really small number and G is an object, called **generator**, we will talk about in a moment.

Such small transformations, when acting on some object change barely anything. In the smallest possible case, such transformations are called **infinitesimal transformations**. However, when we repeat such an infinitesimal transformation many times, we end up with a finite transformation. Think about rotations: Many small rotations in one direction are equivalent to one big rotation in the same direction.

Mathematically, we can write the idea of repeating a small transformation many times as follows

$$h(\theta) = (I + \epsilon G)(I + \epsilon G)(I + \epsilon G)\ldots = (I + \epsilon G)^k, \qquad (3.31)$$

where k denotes how often we repeat the small transformation.

If θ denotes some finite transformation parameter, e.q. 50° or so, and N is some huge number that makes sure we are close to the identity, we can write Eq. (3.30) as

$$g(\theta) = I + \frac{\theta}{N} G. \qquad (3.32)$$

The transformations we want to consider are the smallest possible, which means N must be the biggest possible number, i.e., $N \to \infty$. To get a finite transformation from such an infinitesimal transformation, one has to repeat the infinitesimal transformation infinitely often. Mathematically

$$h(\theta) = \lim_{N \to \infty} (I + \frac{\theta}{N} G)^N, \qquad (3.33)$$

which is in the limit $N \to \infty$ the exponential function[27]

$$h(\theta) = \lim_{N \to \infty} (I + \frac{\theta}{N} G)^N = e^{\theta G}. \qquad (3.34)$$

[27] This is one possible definition of the exponential function. We derive another definition in terms of an infinite series in Appendix A.

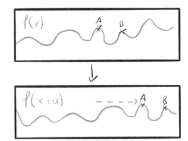

The bottom line is that the object G *generates* the finite transformation h. This is why we call objects like this **generators**.

How do these generators explicitly look like?

Let's consider a function $f(x,t)$ and assume that our goal is to generate a spatial translation $Tf(x,t) = f(x+a,t)$. The generator

$$G_{\text{xtrans}} = \partial_x \qquad (3.35)$$

does the job:[28]

$$e^{aG_{\text{xtrans}}} f(x,t) = (1 + aG_{\text{xtrans}} + \frac{a^2}{2} G_{\text{xtrans}}^2 + \ldots) f(x,t)$$
$$= (1 + a\partial_x + \frac{a^2}{2} \partial_x^2 + \ldots) f(x,t)$$
$$= f(x+a,t)$$

Here we used the series expansion of $e^x = \sum_{n=0}^{\infty} \frac{x^n}{n!}$ and in the last step, we used that in the second to last line we have exactly the Taylor expansion of $f(x+a,t)$.[29] Alternatively consider an infinitesimal translation: $a \to \epsilon$ with $\epsilon \ll 1$. We then have

$$e^{\epsilon G_{\text{xtrans}}} f(x,t) = (1 + \epsilon G_{\text{xtrans}} + \overbrace{\frac{\epsilon^2}{2}}^{\approx 0} G_{\text{xtrans}}^2 + \ldots) f(x,t)$$
$$\approx (1 + \epsilon \partial_x) f(x,t) \qquad \text{(using Eq. 3.35)}$$
$$= f(x,t) + \epsilon \partial_x f(x,t) = f(x+\epsilon, t).$$

Here $\partial_x f(x,t)$ is the **rate of change** of $f(x,t)$ in the x-direction. In other words, this term tells us how much $f(x,t)$ changes when we move in the x-direction. If we multiply this rate of change with the distance that we move in the x-direction - here ϵ - we end up with the total change of $f(x,t)$ if we move by ϵ in the x-direction. Thus, $f(x,t) + \epsilon \partial_x f(x,t)$ really *is* the value of f at the location $x + \epsilon$.

The bottom line is: $G_{\text{xtrans}} = \partial_x$ generates *spatial* translations.

Completely analogous $G_{\text{ttrans}} = \partial_t$ generates *temporal* translations:

$$f(x,t) \to = e^{aG_{\text{ttrans}}} f(x,t) = f(x, t+a).$$

[28] Here ∂_x is a shorthand notation for the derivative $\frac{\partial}{\partial x}$.

[29] If you're unfamiliar with the Taylor series expansion, have a look at Appendix A.

What we have learned here is that generators are the crucial mathematical ingredient that we need to describe continuous symmetries. We can describe *any* continuous symmetry by acting many many times with the corresponding generator on the function in question.

So we can summarize:

> **The core of each continuous symmetry is the corresponding generator.**

This idea is already everything we need to determine *explicitly* how the most important quantum operators look like.

3.4.2 The Most Important Quantum Operators

We now put the puzzle pieces together. The pieces we have so far are:

▷ We are looking for quantum operators. When we act with, for example, the momentum operator on the ket $|\Psi_1\rangle$ that describes our system, we want to get $\hat{p}|\Psi_1\rangle = p_1|\Psi_1\rangle$.

▷ Noether's theorem tells us: there is a deep connection between symmetries and the most important physical quantities: momentum, energy, etc.

▷ Group theory tells us: the essential objects which are responsible for these symmetries are the corresponding generators.

The crucial idea is now to take Noether's theorem seriously. Instead of saying that we get a conserved quantity if there is a symmetry, we say the operator responsible for the symmetry (the generator) is at the same time the operator that describes the conserved quantity (the quantum operator).

[30] Take note that there is an additional imaginary unit i here. This is just a convention that physicists like to use. The translation of a function then works like this: $e^{i\epsilon G_{\text{trans}}} f(x,t)$. So all we have done is introduce two additional i that cancel since $i^2 = -1$. The reason why physicists like to introduce the imaginary unit i here is that we want to interpret the eigenvalues of our quantum operators as something that we can measure in experiments. Without the additional i the eigenvalues would be imaginary. Hence, we introduce an additional i to make them real. However, this is really just a convenient way to make the framework easier to use. In addition, we have an additional minus sign here for the momentum operator. This is motivated by the Minkowski metric of special relativity. However, it is clear that we could also absorb it into the parameter ϵ.

[31] This constant already appeared in the formulas for the relationship between the wavelength and the momentum and energy (Eq. (2.2) and Eq. (2.2)).

[32] To be precise $\hbar$ is the reduced Planck constant (speak: "h-bar") defined by $\frac{h}{2\pi}$ where h is the real Planck constant. The additional factor 2π makes sure that we really get the momentum and energy, as we will see in Section 3.7.

[33] Recall $\partial_x = \frac{\partial}{\partial x}$. The symbol ∂x, in some sense, simply means a tiny amount of x and therefore has the same unit as x. Therefore, $\partial_x \propto 1/\partial x$ has the unit $1/m$.

[34] Square brackets around a quantity means that we are talking about the units of the quantity.

In slogan form:

$$\boxed{\text{quantum operator} \leftrightarrow \text{generator of symmetry}}$$

This may seem like quite a stretch. However, the incredible thing is how well this idea works. We will see this in the next section. But before we move on, let's make the above statement explicit.

Momentum is connected to symmetry under spatial translations. Therefore, we make the identification[30]

$$\boxed{\text{momentum } \hat{p}_i \leftrightarrow \text{generator of spatial translations } -i\hbar\partial_i}$$

(3.36)

Analogously, energy is connected to symmetry under temporal translations. Consequently,

$$\boxed{\text{energy } \hat{E} \leftrightarrow \text{generator of temporal translations } i\hbar\partial_t}$$

(3.37)

Take note that a new fundamental constant appears here: $\hbar$.[31] This constant is known as the **Planck constant**[32] and encodes the magnitude of quantum effects. We can understand the need for a new constant here by observing that momentum has the unit $[p] = \text{kg} \cdot \text{m/s}$. However, ∂_i has the unit[33] $[\partial_i] = 1/\text{m}$. Similarly, ∂_t has the unit $[\partial_t] = 1/\text{s}$, while energy has the unit $[E] = \text{kg} \cdot \text{m}^2/\text{s}^2$. Therefore, we need something to get the same units on the right-hand and left-hand side of the equations. Using the units of energy, momentum and the differential operators we can conclude that $[\hbar] = \text{kg} \cdot \text{m}^2/\text{s}$ since[34]

$$[p] = \text{kg} \cdot \frac{\text{m}}{\text{s}} \stackrel{!}{=} [-i\hbar\partial_i] = \text{kg} \cdot \frac{\text{m}^2}{\text{s}} \frac{1}{\text{m}} = \text{kg} \cdot \frac{\text{m}}{\text{s}} \quad \checkmark$$

$$[E] = \text{kg} \cdot \frac{\text{m}^2}{\text{s}^2} \stackrel{!}{=} [i\hbar\partial_t] = \text{kg} \cdot \frac{\text{m}^2}{\text{s}} \frac{1}{\text{s}} = \text{kg} \cdot \frac{\text{m}^2}{\text{s}^2} \quad \checkmark$$

The Planck constant is one of the most important fundamental constants and we need to extract its value from experiments:

$\hbar \approx 1.055 \cdot 10^{-34}$ kg·m²/s . Since there is no symmetry connected to the conservation of location, the position operator stays what it is[35] $\hat{x}$. In addition, using that angular momentum is defined as $\vec{L} = \vec{x} \times \vec{p}$, we can, in principle, immediately write down the angular momentum operator, simply by replacing $\vec{p}$ with the corresponding quantum operator $\hat{p}$. However, there is an important subtlety, and we postpone the full discussion to Section 3.10.

We are now able to derive one of the most important equations in *all* of physics.

[35] This may seem confusing at first. But all this operator does if we act with it on a function $f(x)$ is to multiply it with x, i.e., $\hat{x} f(x) = x f(x)$ since x really is the location we evaluate $f(x)$ at.

3.5 How Operators Influence Each Other

As already emphasized above, the crucial new thing in Quantum Mechanics is that there is a fundamental uncertainty. This uncertainty comes about since each time we measure the location we change the momentum, and each time we measure the momentum we change the location. Mathematically this statement means that it makes a difference whether we first measure the momentum or first measure the location: $\hat{x}\hat{p}\,|\Psi\rangle \neq \hat{p}\hat{x}\,|\Psi\rangle$. We can also write this statement as $(\hat{x}\hat{p} - \hat{p}\hat{x})\,|\Psi\rangle \neq 0$. The shorthand notation for $AB - BA$ is $[A, B]$ and we call this usually the **commutator** of A and B.

To get an intuitive understanding for what it means that two operators do not commute, compare the situation where you first put on your socks and then your shoes with the situation where you first put on your shoes and then your socks. The outcome is clearly different and the ordering of the operations "putting shoes on" and "putting socks on" therefore matters. In technical terms, we say these two operations do not commute. In contrast, it makes no difference if you first put on your left sock and then your right sock or first your right sock and then your left sock. These two operations do commute.

Now, using the explicit quantum operator $\hat{p}_i = -i\partial_i$ (Eq. (3.36)) we can actually derive that $[\hat{p}_i, \hat{x}_j] \neq 0$:

$$\begin{aligned}
[\hat{p}_i, \hat{x}_j]\,|\Psi\rangle &= (\hat{p}_i\hat{x}_j - \hat{x}_j\hat{p}_i)\,|\Psi\rangle \\
&= (-i\hbar\partial_i \hat{x}_j + \hat{x}_j i\hbar\partial_i)\,|\Psi\rangle && \text{(Eq. 3.36)} \\
&= -(i\hbar\partial_i \hat{x}_j)\,|\Psi\rangle - \hat{x}_j(i\hbar\partial_i|\Psi\rangle) + \hat{x}_j i\hbar\partial_i|\Psi\rangle && \text{(product rule)} \\
&= -i\hbar\delta_{ij}\,|\Psi\rangle && (\partial_i \hat{x}_j = \delta_{ij})
\end{aligned}$$

[36] The Kronecker delta that appears here δ_{ij} is zero for $i \neq j$ and one for $i = j$. If you're unfamiliar with it, have a look at Appendix C.

In the last step we used that, for example, $\partial_y x = 0$ but $\partial_y x = 1$.[36] We didn't assume anything about $|\Psi\rangle$ so the equation is valid for any $|\Psi\rangle$. Therefore, we can write the equation a without it which makes the equation a bit shorter. However, we always have to remember that there is an implicit ket in such equations and we are only too lazy to write it all the time. In conclusion:

$$[\hat{p}_i, \hat{x}_j] = -i\hbar \delta_{ij} \quad (3.38)$$

This little equation is known as the **canonical commutation relation**. As the name already indicates this equation is extremely important. In fact, many textbooks and lectures use it as a *starting point* for Quantum Mechanics.[37]

The canonical commutation relation encodes an incredibly important property of Quantum Mechanics: It makes a difference in which order we measure the location and the momentum! This is exactly the quantum uncertainty we talked about previously[38]. In physical terms, it means that we can't measure the position and the momentum of a particle at the same time with arbitrary precision. We can now also understand how this comes about in our framework. We identified the momentum operator with the generator of spatial translations (Eq. (3.36)). Thus, each time we measure the momentum, we perform a tiny spatial translation. What we do here is therefore analogous to the example we discussed in Section 2.3 where we argued that each time we measure the location of an electron we have to push it a little with a photon.

The next cool thing we can do using the explicit quantum operators is to derive how quantum systems change as time passes on. This is described by the second most important equation of Quantum Mechanics, which is what we talk about next.

3.6 How Quantum Systems Change

So far, we acted as if quantum systems were static. However, in the real world, things are usually subject to constant change. Therefore, we need something that allows us to understand how

[37] Formulated differently, many textbooks use this equation as the fundamental postulate of Quantum Mechanics. For example, using this equation, we can derive how the momentum operator looks like etc.

[38] We talked about the quantum uncertainty in Section 2.3. The connection between the commutation relation and uncertainty can be made mathematically precise. Using mathematical results such as the Cauchy-Schwartz inequality and the explicit definition of the standard deviation it is possible to derive for two general operators $\hat{A}$ and $\hat{B}$ the following equation

$$(\Delta \hat{A})^2 (\Delta \hat{B})^2 \geq |\frac{1}{2i}\langle [\hat{A}, \hat{B}] \rangle|,$$

where $\Delta \hat{A}$ and $\Delta \hat{B}$ are the standard deviations of the two operators. Therefore, using Eq. (3.38) we find

$$(\Delta \hat{p})^2 (\Delta \hat{x})^2 \geq |\frac{1}{2i}\langle [\hat{p}, \hat{x}] \rangle| = \frac{\hbar}{2}.$$

A measurement without any variance would mean $\Delta \hat{p} = 0$ plus $\Delta \hat{x} = 0$. This equation tells us that this is impossible. It's impossible to measure the location and the momentum of a particle at the same time with arbitrary precision. If we reduce, for example, the variance in our location measurements we automatically get a larger variance in our momentum measurements.

a quantum system evolves as time passes on. Mathematically, what we need is an equation that looks like this

$$\partial_t |\Psi\rangle = \ldots$$

Here, $\partial_t |\Psi\rangle$ is the rate of change of $|\Psi\rangle$ if we move in the t-direction. In other words, $\partial_t |\Psi\rangle$ describes how $|\Psi\rangle$ changes as time passes on and is therefore exactly what we need. On the right-hand side, we need something that contains specific details about the system in question. Thus, the right-hand side will be different for different systems. Then, as soon as we have such an equation, we have to solve it to understand how $|\Psi\rangle$ depends on t.

Luckily, we don't have to guess. All we have to do is take a second look at what we already derived in the last section. The main actor here: ∂_t is almost exactly the quantum energy operator $i\hbar\partial_t$ (Eq. (3.37)), only without the imaginary unit and Planck's constant. This means we now know what we have on the right-hand side: the energy:

$$i\hbar\partial_t |\Psi\rangle = E |\Psi\rangle \, .$$

Is there anything else we know about energy?

In Classical Mechanics, the total energy is given as the sum of kinetic energy and potential energy

$$E = T + V \equiv \text{kinetic energy } + \text{ potential energy.} \quad (3.39)$$

The usual formula for the kinetic energy is

$$T = \frac{1}{2}mv^2 = \frac{p^2}{2m} \quad (3.40)$$

since $p = mv$. We can turn this equation into a quantum equation, simply by replacing p with the quantum operator $\hat{p}_i = -i\hbar\partial_i$ that we derived above[39]. The potential energy usually only depends on the position: $V = V(x)$. Therefore, we can turn it into a quantum operator by replacing $x \to \hat{x}$.[40]

So what we can do now is go back to our equation

$$i\hbar\partial_t |\Psi\rangle = E |\Psi\rangle,$$

[39] The index i appears here since in general, we are dealing with systems that can change in 3-spatial dimension. The index takes on the values $i = \{x, y, z\}$ or equivalently $i = \{1, 2, 3\}$.

[40] If there is more than one dimension, we have $\hat{x}_i$.

use on the right-hand side the usual formula $E = T + V$ and replace all quantities that appear here with the corresponding quantum operators:

$$\begin{aligned}i\hbar\partial_t \left|\Psi\right\rangle &= E\left|\Psi\right\rangle \\ &= (T+V)\left|\Psi\right\rangle &\text{(using Eq. 3.39)} \\ &= \left(\frac{p^2}{2m} + V(x)\right)\left|\Psi\right\rangle &\text{(using Eq. 3.40)} \\ \Rightarrow \quad i\hbar\partial_t \left|\Psi\right\rangle &= \left(\frac{\hat{p}^2}{2m} + V(\hat{x})\right)\left|\Psi\right\rangle &\text{(introducing operators)} \\ &= \left(\frac{(-i\hbar\partial_i)^2}{2m} + V(\hat{x})\right)\left|\Psi\right\rangle &\text{(using Eq. 3.36)} \\ &= \left(-\frac{\hbar^2\partial_i^2}{2m} + V(\hat{x})\right)\left|\Psi\right\rangle . \end{aligned}$$

The resulting equation

$$i\hbar\partial_t \left|\Psi\right\rangle = -\frac{\hbar^2\partial_i^2}{2m}\left|\Psi\right\rangle + V(\hat{x})\left|\Psi\right\rangle \qquad (3.41)$$

is the famous **Schrödinger equation**. It describes correctly how quantum systems evolve in time. Most problems in Quantum Mechanics consist of solving the Schrödinger equation for different potentials $V(x)$ and different boundary conditions. We will discuss the most important systems in Chapter 5. For historical reasons the energy operator that appears on the right-hand side of the Schrödinger equation is known as the **Hamiltonian operator**

$$\hat{H} \equiv \left(-\frac{\hbar^2\partial_i^2}{2m} + V(\hat{x})\right) . \qquad (3.42)$$

In the more general form

$$i\hbar\partial_t \left|\Psi\right\rangle = \hat{H}\left|\Psi\right\rangle \qquad (3.43)$$

the Schrödinger equation is also valid for relativistic systems and even in Quantum Field Theory. All we have to do is modify $\hat{H}$ appropriately[41].

A convenient alternative way to describe the time-evolution in Quantum Mechanics is with the so-called **time evolution**

[41] In fact, we can simply use the relativistic energy-momentum relation $E^2 = (pc)^2 + (mc^2)^2$ instead of the non-relativistic one $E = \frac{p^2}{2m}$. Then again we replace p with the corresponding quantum operator and end up with an equation that describes how relativistic quantum system evolve in time. The resulting equation is known as Klein-Gordon equation and is the correct equation of motion for (spinless) relativistic particles.

operator $U(t)$. We define this operator through the following property

$$|\Psi(x,t)\rangle = U(t)|\Psi(x,0)\rangle \tag{3.44}$$

In words, this means that if we act with this operator on some ket $|\Psi(x,0)\rangle$, the result is the ket that describes the system at a later point in time t: $|\Psi(x,t)\rangle$. This operator is not merely an abstract thing. We can write it down explicitly by putting Eq. (3.44) into our Schrödinger equation:

$$i\hbar\partial_t|\Psi(x,t)\rangle = H|\Psi(x,t)\rangle$$
$$\therefore \quad i\hbar\partial_t U(t)|\Psi(x,0)\rangle = HU(t)|\Psi(x,0)\rangle \quad \text{(using Eq. 3.44)}$$

This equation holds for any $|\Psi(x,0)\rangle$ and we can therefore write it without it:

$$\therefore \quad i\hbar\partial_t U(t) = HU(t)$$
$$\therefore \quad i\hbar\frac{\partial_t U(t)}{U(t)} = H.$$

This is a differential equation for $U(t)$ and the general solution is

$$U(t) = e^{-\frac{i}{\hbar}\int_0^t dt' H(t')} \tag{3.45}$$

since

$$i\hbar\frac{\partial_t U(t)}{U(t)} = H$$

$$\therefore \quad i\hbar\frac{\partial_t e^{-\frac{i}{\hbar}\int_0^t dt' H(t')}}{e^{-\frac{i}{\hbar}\int_0^t dt' H(t')}} = H \quad \text{(using Eq. 3.45)}$$

$$\therefore \quad i\hbar\left(-\frac{i}{\hbar}\underbrace{\partial_t \int_0^t dt' H(t')}_{=H}\right)\frac{\cancel{e^{-\frac{i}{\hbar}\int_0^t dt' H(t')}}}{\cancel{e^{-\frac{i}{\hbar}\int_0^t dt' H(t')}}} = H$$

$$\therefore \quad H = H \quad \checkmark$$

So the time-evolution operator is simply a convenient way to write the information encoded in the Schrödinger equation a bit differently.

We can now answer one question that was left open until now: Where in the quantum framework can we see that Quantum Mechanics is about waves?

3.7 Why Quantum Mechanics is About Waves

Let's consider a quantum system without any potential $V(\hat{x}) = 0$. The time evolution of the system is described by the equation

$$\text{Eq. (3.41) with } V = 0 \quad \leftrightarrow \quad i\hbar\partial_t |\Psi\rangle = -\frac{\hbar^2 \partial_i^2}{2m} |\Psi\rangle$$

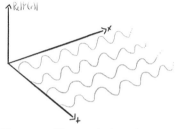

Figure 3.1: Time evolution of a plane wave. Since, in general, a wave function is complex and therefore hard to visualize, here only the real part is shown.

Before we solve this equation, to make things simpler, we choose a specific basis. We use the position basis and therefore the wave functions that we introduced in Section 3.3. Also, to unclutter the notation, we restrict ourselves to one spatial dimension. Explicitly, we then have

$$i\hbar\partial_t |\Psi\rangle = -\frac{\hbar^2 \partial_x^2}{2m} |\Psi\rangle$$

$$\underset{\text{Eq. (3.24)}}{\therefore} \quad i\hbar\partial_t \int dx \Psi(x,t) |x\rangle = -\frac{\hbar^2 \partial_x^2}{2m} \int dx \Psi(x,t) |x\rangle$$

$$\therefore \quad i\hbar\partial_t \Psi(x,t) = -\frac{\hbar^2 \partial_x^2}{2m} \Psi(x,t).$$

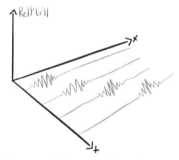

Figure 3.2: Time evolution of a wave packet. Again, only the real part is shown.

This is a legitimate approach, as long as we remember that $\Psi(x,t)$ is defined under an integral. Hence, it describes a *density*. Since the coefficients in front of the kets are probability amplitudes, the wave function $\Psi(x,t)$ is a **probability amplitude density**. To get probabilities[42], we have to take the absolute square and then integrate it over some spatial region. The result we get this way is the probability to find the particle in the region that we integrated over. Now a solution of this equation is

[42] After all, probabilities are what we really can measure in experiments.

$$\Psi(x,t) = e^{-i(Et-px)/\hbar} \qquad (3.46)$$

since

$$i\hbar \partial_t \Psi(x,t) = -\frac{\hbar^2 \partial_x^2}{2m}\Psi(x,t)$$

$$\therefore i\hbar \partial_t e^{-i(Et-px)/\hbar} = -\frac{\hbar^2 \partial_x^2}{2m} e^{-i(Et-px)/\hbar} \quad \text{(using Eq. 3.46)}$$

$$\therefore E e^{-i(Et-px)/\hbar} = \frac{p^2}{2m} e^{-i(Et-px)/\hbar}$$

$$\therefore \frac{p^2}{2m} e^{-i(Et-px)/\hbar} = \frac{p^2}{2m} e^{-i(Et-px)/\hbar} \checkmark, \quad (3.47)$$

where in the last step we used that E is just the numerical value for the total energy of a free particle: $E = \frac{p^2}{2m}$. A function of the form $e^{-i(Et-\vec{p}\vec{x})/\hbar}$ is known as a **plane wave**[43].

An important observation is that the Schrödinger equation is linear in Ψ. This means that we don't have any terms like ψ^2, ψ^3 etc. A direct consequence of this property is that if we have two or more solutions ψ_1, ψ_2, ..., we can immediately write down additional solutions

$$\psi_{\text{sup}} = a\psi_1 + b\psi_2 + \ldots \quad (3.48)$$

This is known as a **superposition**. We can see that a superposition is also a solution by putting it into the Schrödinger equation.:

$$H\psi_{\text{sup}} = i\hbar \partial_t \psi_{\text{sup}} \quad \text{(this is Eq. 3.43)}$$

$$\therefore H(a\psi_1 + b\psi_2 + \ldots) = \partial_t(a\psi_1 + b\psi_2 + \ldots) \quad \text{(using Eq. 3.48)}$$

The functions ψ_1, ψ_2 etc. are solutions, which means

$$H\psi_1 = E_1 \psi_1, \quad H\psi_2 = E_2 \psi_2 \quad (3.49)$$

and

$$i\hbar \partial_t \psi_1 = E_1 \psi_1 \quad , \quad i\hbar \partial_t \psi_2 = E_2 \psi_2. \quad (3.50)$$

Putting this into our equation above for the superposition yields

$$H(a\psi_1 + b\psi_2 + \ldots) = i\hbar \partial_t(a\psi_1 + b\psi_2 + \ldots)$$
$$\to aH\psi_1 + bH\psi_2 + \ldots = ai\hbar \partial_t \psi_1 + bi\hbar \partial_t \psi_2 + \ldots$$
$$aE_1\psi_1 + bE_2\psi_2 + \ldots = aE_1\psi_1 + bE_2\psi_2 + \ldots \checkmark$$

[43] If you wonder why this is the case, remember Euler's formula: $e^{ix} = \cos(x) + i\sin(x)$. The well-known $\cos(x)$ and $\sin(x)$ are oscillating functions that perfectly describe waves. In other words, the real and imaginary parts of a function that looks like this e^{ix} describes an oscillation that spreads out all over space with equal amplitude. (Hence the prefix *plane wave*.)

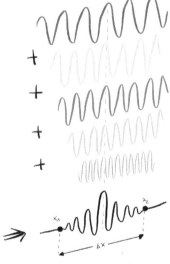

Figure 3.3: Construction of a wave packets as a superposition of plane waves.

So the linear combination $\psi_{\text{sup}} = a\psi_1 + b\psi_2 + \ldots$ is really also a solution[44].

This observation is important since we can construct **wave packets** through suitable linear combinations of plane waves.

One possibility is a Gaussian wave-packet, where $A(\vec{p})$ is a Gauss distribution[45]

$$\Psi_{GWP}(\vec{x},t) = \int dp^3 A(\vec{p}) e^{i(\vec{p}\vec{x}-Et)/\hbar}$$
$$= \int dp^3 \psi_0 e^{i(\vec{p}-\vec{p})^2/4\sigma^2} e^{i(\vec{p}\vec{x}-Et)/\hbar} .$$

You can see an example of such a Gaussian wave-packet in the margin.

A wave packet is what we use to describe a particle which is localized within some region. Localized within some region means that we have some idea where the particle is. Without any information the particle could be anywhere and hence the corresponding wave function is not localized but spreads out all over space. A plane wave alone is not useful to describe a real particle since it spreads out all over space with equal amplitude. Only suitable linear combinations of plane waves are useful to describe real particles.

Take note that solutions of the Schrödinger equation are *complex* functions. Therefore, as promised above, we can now understand that physical kets are in general complex[46] and this explains why we need to additionally complex conjugate them to construct the corresponding bras. In addition, it is always important to keep in mind that solutions of the Schrödinger equation only represent probability amplitudes. To get probabilities we have to take the absolute square of them.

Another thing we can now understand is the De Broglie relations (Eq. (2.2), Eq. (2.3)).[47] In general, a plane wave has the form: $Ae^{i\frac{2\pi x}{\lambda} - \omega t}$, where λ is the wavelength and ω the angular frequency of the wave. Comparing this with our solution of the Schrödinger equation in Eq. (3.46): $\Psi = e^{i(px-Et)/\hbar}$ tells us that[48]

[44] Again: This only works because there is no term ψ^2 or ψ^3. In other words, because the Schrödinger equation is linear in ψ. If you don't believe this, try to do the same calculation with an additional term ψ^2 or ψ^3 in the Schrödinger equation.

[45] Again, we imagine that an integral is something like a finely grained sum.

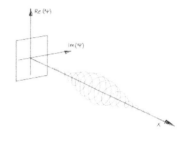

[46] A physical ket is one that is a solution of the Schrödinger equation.

[47] Reminder: $\lambda = \frac{h}{p}$ and $E = h\nu$.

[48] Recall that $\hbar$ is the reduced Planck constant defined by $\hbar = h/2\pi$. Here we can understand why we introduced this new constant with an additional factor 2π.

$$\frac{2\pi}{\lambda} \hat{=} \frac{p}{\hbar}$$
$$\therefore \quad p = \frac{\hbar 2\pi}{\lambda} = \frac{h}{\lambda}$$

and we also find

$$\omega \hat{=} \frac{E}{\hbar}$$
$$\therefore \quad E = \hbar\omega = h\nu,$$

where ν is the "ordinary" frequency of the wave (in contrast to the angular frequency $\omega = \nu/2\pi$).

Before we move on, we should talk about one concept which we have already used several times, in more detail.

3.8 Intermezzo: Eigenvectors and Eigenvalues

Two notions that are used in Quantum Mechanics all the time are: **eigenvalue** and **eigenvector**. The eigenvectors $\vec{v}$ and eigenvalues λ for a matrix M are exactly those vectors and numbers that fulfill the equation

$$M\vec{v} = \lambda\vec{v}. \tag{3.51}$$

The important thing is that we have *exactly* the same vector $\vec{v}$ on both sides. This means that the vector $\vec{v}$ remains, up to a constant λ, unchanged if we multiply it with the matrix M. For each eigenvector we have a corresponding eigenvalue.

Now this is just a definition. Why should we care about eigenvalues and eigenvectors?

Two reasons:

▷ Think about rotations. We can describe rotations with rotation matrices and the eigenvector of a given rotation matrix is exactly the rotational axis.

▷ The structure operator × vector = number × same vector is *exactly* the structure that we use in Quantum Mechanics. A matrix is an operator just like the quantum operators we talked about above[49]. The bottom line here is that the eigenvectors are the states with definite values for the physical quantity represented by the operator. For example, if $|\Psi_1\rangle$ is an eigenvector of the momentum operator $\hat{p}$ we have: $\hat{p}|\Psi_1\rangle = p_1|\Psi_1\rangle$.[50] The corresponding eigenvalue p_1 is the momentum value that we measure. Many operators in Quantum Mechanics are not represented by matrices but by differential operators[51]. However, the structure $\hat{O}|\Psi\rangle = o|\Psi\rangle$ is completely analogous. The only thing that changes if $\hat{O}$ is a differential operator is that people talk about **eigenfunctions** instead of eigenvectors.

[49] In fact, in the next section we will talk about a quantum operator that is indeed represented by a matrix.

[50] Take note that a general ket is not an eigenvector of the momentum operator. For example

$$|\Psi\rangle = a|\Psi_1\rangle + b|\Psi_2\rangle$$
$$\Rightarrow \hat{p}|\Psi\rangle = a\hat{p}|\Psi_1\rangle + b\hat{p}|\Psi_2\rangle$$
$$= ap_1|\Psi_1\rangle + bp_2|\Psi_2\rangle$$
$$\neq c(a|\Psi_1\rangle + b|\Psi_2\rangle).$$

The crucial observation here is that we don't have the same vector on the left-hand and right-hand side since we can't write $ap_1|\Psi_1\rangle + bp_2|\Psi_2\rangle$ as some number times the original vector $(a|\Psi_1\rangle + b|\Psi_2\rangle)$ since, in general, $p_1 \neq p_2$.

There are sophisticated algorithms to calculate the eigenvectors and eigenvalues of a matrix or the eigenfunctions of a differential operator. But a full discussion would lead us too far astray. You can find a full description of these algorithms (at least for matrices) in any linear algebra textbook.[52]

Now it's finally time to return to the angular momentum operator, as promised in Section 3.4.

[51] Think: something with a derivative in it ∂_x. Examples are the momentum operator $-i\hbar\partial_x$ or the energy operator $i\hbar\partial_t$.

[52] The part of mathematics that deals specifically with these kinds of problems is known as **spectral theory**.

3.9 Angular Momentum

In Section 3.4.2 I already mentioned that the angular momentum operator follows directly from the classical angular momentum $\vec{L} = \vec{x} \times \vec{p}$ if we replace $\vec{p}$ with the corresponding momentum operator $-i\hbar\vec{\partial}$.

In addition, we learned in Section 3.5 that it makes a difference whether we first measure the location and then the momentum or first the momentum and then the location. This curious property of Quantum Mechanics is encoded in the canonical commutation relation (Eq. (3.38))[53].

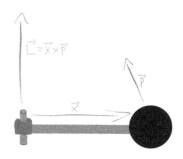

[53] Reminder: $[\hat{p}_i, \hat{x}_j] = -i\hbar\delta_{ij}$

Now the thing is that we can find a very similar relation for the angular momentum operator. Using the explicit form of the angular momentum operator[54]

$$\begin{aligned}\hat{L}_i &= \epsilon_{ijk}\hat{x}_j\hat{p}_k \\ &= \epsilon_{ijk}\hat{x}_j(-i\hbar\partial_k) \quad \text{(using Eq. 3.36)} \\ &= -i\hbar\epsilon_{ijk}\hat{x}_j\partial_k\end{aligned}$$

[54] It is convenient to write the operator in index notation. The cross product $\vec{A} \times \vec{B}$ reads in index notation $\epsilon_{ijk}A_iB_j$, where ϵ_{ijk} is the Levi-Civita symbol. For the definition, see Eq. 2.

we can calculate[55]

$$[\hat{L}_i, \hat{L}_j] = i\hbar\epsilon_{ijk}\hat{L}_k \quad (3.52)$$

[55] Explicitly this equation reads, for example, $[\hat{L}_x, \hat{L}_y] = i\hbar\hat{L}_z$ and $[\hat{L}_y, \hat{L}_z] = i\hbar\hat{L}_x$ and $[\hat{L}_z, \hat{L}_x] = i\hbar\hat{L}_y$. To calculate it you have to remember that there is an implicit ket $|\Psi\rangle$ behind each term which we are too lazy to write all the time.

This equation is known as the **angular momentum commutation relation** or also **angular momentum algebra**. Take note that the crucial point here is that measurements of the angular momentum along *different* axis affect each other. So concretely, a measurement of the angular momentum along the x-axis affects the angular momentum along the y- and z-axis. This is in contrast to the momentum and location operators, where measurements along *different* axes do not affect each other since, for example, $\partial_x y = 0$ and $\partial_x \partial_y = \partial_y \partial_x$.

In practice, this means that we can't determine the angular momentum of a given particle along different axes with arbitrary precision. Each time we measure it along one axis, we lose information about the angular momentum along the other axis.

However the craziness of angular momentum in Quantum Mechanics does not stop here. Next, we talk about an important subtlety of the angular momentum operator in Quantum Mechanics. As we will see in a moment, this subtlety leads us to a new and incredibly important quantum quantity. Unfortunately, the full story is far too long to include it here. If you are interested, you can find all the details in my other book[56]. The following section will only sketch the full story.

[56] Jakob Schwichtenberg. *Physics from Symmetry*. Springer, Cham, Switzerland, 2018. ISBN 978-3319666303

3.10 Spin

For many systems, we can use the "naive" operator that we get by using $\vec{L} = \vec{x} \times \vec{p}$ and then replacing $\vec{p}$ with $-i\hbar\vec{\partial}$. The resulting operator is exactly the generator of rotations. For example, let's say we have a function $f(x)$ and want to describe it in a rotated coordinate system. We can achieve an infinitesimal rotation as follows:

$$f(x) \to (1 + \epsilon \hat{L}) f(x).$$

However, this is not always the full story.

Sometimes, it's necessary that we describe our quantum system not just with a function $\Psi(x,t)$, but with objects that have more than one component[57]

$$\Psi(x,t) = \begin{pmatrix} \Psi_1(x,t) \\ \Psi_2(x,t) \end{pmatrix}.$$

A crucial point is that each component of these objects is a *function* of x and t. This means that the action of the symmetry generators on such objects do two things[58]

1. On the one hand, they change the spatial and time coordinates $x \to 1 + Gx$, where G denotes a generator. This is exactly the kind of transformation that we talked about so far: $\Psi(x) \to \Psi(x + \epsilon)$

2. But on the hand, in general, a transformation of an object with several components can also mix these components:
$$\begin{pmatrix} \Psi_1(x,t) \\ \Psi_2(x,t) \end{pmatrix} \to \begin{pmatrix} \Psi_2(x,t) \\ \Psi_1(x,t) \end{pmatrix}$$

The most familiar example is when we rotate a vector. Under a rotation, the vector components get mixed[59]. And when the components depend on the position x, these functions get modified, too.

---

[57] Think: like a vector. The reason why we need such objects with more than one component is that they are necessary to describe certain quantum systems. For example, you maybe already know that electromagnetic waves can be polarized. This means there are internal degrees of freedom and we therefore need objects with more than one component to describe them. As already mentioned above, electromagnetic waves are transmitted by particles (photons) and therefore we need objects with several components to describe these quantum particles. We will talk about another example in a moment.

[58] For the other operators there is no such subtlety since translations do not mix components.

[59] An example: We want to look at a vector with four components $A_\mu = \begin{pmatrix} A_0 \\ A_1 \\ A_2 \\ A_3 \end{pmatrix}$ from a different perspective. Formulated differently, we want to describe it in a rotated coordinate system. The result could be something like $A'_\mu = \begin{pmatrix} A'_0 \\ A'_1 \\ A'_2 \\ A'_3 \end{pmatrix} = \begin{pmatrix} A_0 \\ -A_2 \\ A_1 \\ A_3 \end{pmatrix}$. Here A'_μ and A_μ describe the same field in coordinate systems that are rotated by 90° around the z-axis relative to each other. Through the transformation the components get mixed.

[60] Think: something with a derivative ∂_i in it.

[61] Orbital angular momentum is an important quantity, for example, when an object revolves around a second one. Think: earth and the sun.

[62] Think: a spinning ball, although you shouldn't take this picture too seriously. This is in contrast to orbital angular momentum which only exists when two objects revolve around each other.

[63] The correct name for the two-component objects that we deal with in Quantum Mechanics is **spinors**. Spinors are somewhat strange objects and not simply two-dimensional vectors. However, a full discussion and a derivation of these generators is too long to include it here. If you're interested you can you can find all the details, for example, in my other "Physics from Symmetry".

[64] Here we use the labels 1, 2, and 3 instead of x, y, and z. However, whenever we talk about, for example the 3 axis, you can equally imagine that we talk about the z-axis.

The generator of rotations that causes the mixing of the components is a *matrix*. The generator that causes the rotation of the argument of a function is a *differential operator*[60].

Our goal is to find the general quantum operator that describes angular momentum. For momentum and energy, the identification of quantum operators with generators was simple. However, now we have a generator that consists of two parts. Which part is the correct one?

Both! For the angular momentum that we know from Classical Mechanics, the naive operator $\vec{L} = \vec{x} \times (-i\hbar\vec{\partial})$ is correct. This type of angular momentum is known as **orbital angular momentum**.[61]

However, the second operator is equally important. We identify it with the generator that causes the mixing of the components (i.e., is a matrix). Since both quantities follow when we consider rotations they can't be too different. And in fact, the quantity that is described by this second operator is commonly interpreted as *internal* angular momentum[62]. We call this internal angular momentum usually **spin**. The total angular momentum is the sum of these two quantities.

The correct generators for two-component objects are[63]

$$\hat{S}_i = \frac{\hbar}{2}\sigma_i \qquad (3.53)$$

where σ_i are (2×2) matrices known as **Pauli matrices**:[64]

$$\sigma_1 = \begin{pmatrix} 0 & 1 \\ 1 & 0 \end{pmatrix}, \quad \sigma_2 = \begin{pmatrix} 0 & -i \\ i & 0 \end{pmatrix}, \quad \sigma_3 = \begin{pmatrix} 1 & 0 \\ 0 & -1 \end{pmatrix}.$$

For example, the operator $\hat{S}_3$ reads explicitly

$$\hat{S}_3 = \frac{\hbar}{2}\sigma_3 = \begin{pmatrix} \frac{\hbar}{2} & 0 \\ 0 & -\frac{\hbar}{2} \end{pmatrix}. \qquad (3.54)$$

The corresponding **eigenvectors** are

$$v_{\frac{\hbar}{2}} = \begin{pmatrix} 1 \\ 0 \end{pmatrix} \qquad v_{-\frac{\hbar}{2}} = \begin{pmatrix} 0 \\ 1 \end{pmatrix} \qquad (3.55)$$

with **eigenvalues** $\frac{\hbar}{2}$ and $-\frac{\hbar}{2}$, respectively. These eigenvectors are our basic spin states. The eigenvalues are the values that we can measure. Let's say our system is in a state described by

$$\Psi(x,t) = \begin{pmatrix} \psi(x,t) \\ 0 \end{pmatrix}. \quad (3.56)$$

We want to measure the spin in the 3-direction. Therefore, we calculate

$$\hat{S}_3 \Psi(x,t) = \hat{S}_3 \begin{pmatrix} \psi(x,t) \\ 0 \end{pmatrix} \quad \text{(using Eq. 3.56)}$$

$$= \begin{pmatrix} \frac{\hbar}{2} & 0 \\ 0 & -\frac{\hbar}{2} \end{pmatrix} \begin{pmatrix} \psi(x,t) \\ 0 \end{pmatrix} \quad \text{(using Eq. 3.54)}$$

$$= \frac{\hbar}{2} \begin{pmatrix} \psi(x,t) \\ 0 \end{pmatrix}.$$

This is exactly the structure we talked about in the sections above. We act with the quantum operator on the object that describes our state and get back what we will measure in an experiment. In this case we would measure the value $\hbar/2$ for the spin in the 3-direction[65].

[65] The labels do not really matter. We can equally call our 3-direction the z-direction.

Again, in a general a system will be in a state that is a mixture of different states with specific spin. A convenient notation is $|\uparrow\rangle$ for the state with spin $\hbar/2$ and $|\downarrow\rangle$ for the state with spin $-\hbar/2$.[66] A general state therefore reads

[66] The state with spin $\hbar/2$ is usually called "spin up" state and the state with spin $-\hbar/2$ "spin down" state.

$$|\Psi\rangle = a|\uparrow\rangle + b|\downarrow\rangle.$$

The coefficients a and b are again directly related to the probabilities to measure $\hbar/2$ or $-\hbar/2$.[67]

[67] Explicitly: $|a|^2$ is the probability to measure $\hbar/2$ and $|b|^2$ the probability to measure $-\hbar/2$.

Now comes the crucial point.

While in Classical Mechanics, we can measure any value for the angular momentum, for this new type of angular momentum this is no longer possible. Instead, there are only two possibilities. Either we measure the value $\hbar/2$, or we measure the value $-\hbar/2$. There is nothing in between[68]. In other words:

[68] However, for orbital angular momentum in most cases still, any value is possible.

> **Spin is quantized!**

Historically this was a huge surprise for every physicist. The quantization of angular momentum was only discovered experimentally and not predicted by any theorist. The discovery was made by the second most famous quantum experiment[69]: the **Stern-Gerlach experiment**. The experimental setup is shown in the following image[70]

[69] The most famous quantum experiment is the double slit experiment that we already talked about in Section 2.1.

[70] Don't let yourself get confused by the fact that historically silver atoms were used. They did this solely for technical reasons. The angular momentum of silver atoms depends dominantly on the angular momentum of a single electron. Since an experiment with single electrons was too difficult at the time, this was the next best thing they could do. Nowadays, physicists are able to repeat the experiment with single electrons. The result is the same.

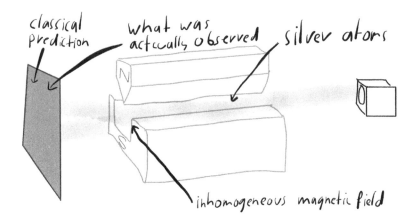

Nowadays, with the power of hindsight, this result is no longer completely mysterious. As we have seen above, this result can be derived completely analogous to all other quantum features and is not really something completely new.

However, this is not the only curious property of spin. Completely analogous to what we discussed for "ordinary" angular momentum, we can now check if spin measurements along different axes affect each other. We already learned that measurements of momentum and location regarding the *same* axis affect each other. And also we already learned that measurements of angular momentum along *different* axis affect each other. Mathematically, these properties are encoded in the canonical commutation relation (Eq. (3.38))[71] and the angular momentum commutation relation (Eq. (3.52))[72].

Using the explicit matrices for the spin operators (Eq. 3.53) we can calculate[73]

$$[\hat{S}_i, \hat{S}_j] = i\hbar \epsilon_{ijk} \hat{S}_k. \quad (3.57)$$

This commutation relation is known as the **spin algebra**. It tells us, for example, that it makes a difference whether we first measure the spin in the x direction and then in the y or the other way round. Take note that this relation is completely analogous to the angular momentum algebra which once more tells us that spin is some type of angular momentum, too. We will talk a bit more about spin in Section 10.1.

An important side-note is that not every object has spin. However, almost all elementary particles have a nonzero spin. The only exception is the recently observed Higgs boson. For electrons (and also, for example, protons or neutrons or quarks) we need to use two-component objects. We say they have spin 1/2, for reasons that we will discuss in more detail in Section 10.1. Other elementary particles like, for example, photons have spin 1, and we need four-component objects (four-vectors) to describe them. The corresponding spin operators that allow us to extract the spin values of photons are therefore (4×4) matrices.

So what we have learned in the previous sections is that in Quantum Mechanics it's impossible to know the location and the momentum at the same time with arbitrary precision. Moreover, we learned that we can't know the full angular momentum vector $\vec{L} = (L_x, L_y, L_z)^T$ since each time we measure it along one axis, we lose information about the angular momentum along the other axes[74].

Thus, we have to be a bit more careful in Quantum Mechanics

[71] Reminder: $[\hat{p}_i, \hat{x}_j] = -i\hbar \delta_{ij}$ and the square brackets denote in general commutators: $[A, B] = AB - BA$. A commutator encodes if it makes a difference if we first apply B and the A or first A and then B. Formulated differently, a nonzero commutator indicates that it makes a difference in what order we apply two operators.

[72] Reminder: $[\hat{L}_i, \hat{L}_j] = i\hbar \epsilon_{ijk} \hat{L}_k$.

[73] For example, we have

$$[\hat{S}_1, \hat{S}_2] = \hat{S}_1 \hat{S}_2 - \hat{S}_2 \hat{S}_1$$
$$= \begin{pmatrix} 0 & \frac{\hbar}{2} \\ \frac{\hbar}{2} & 0 \end{pmatrix} \begin{pmatrix} 0 & -i\frac{\hbar}{2} \\ i\frac{\hbar}{2} & 0 \end{pmatrix}$$
$$- \begin{pmatrix} 0 & -i\frac{\hbar}{2} \\ i\frac{\hbar}{2} & 0 \end{pmatrix} \begin{pmatrix} 0 & \frac{\hbar}{2} \\ \frac{\hbar}{2} & 0 \end{pmatrix}$$
$$= \begin{pmatrix} \frac{i\hbar^2}{4} & 0 \\ 0 & -\frac{i\hbar^2}{4} \end{pmatrix} - \begin{pmatrix} -\frac{i\hbar^2}{4} & 0 \\ 0 & \frac{i\hbar^2}{4} \end{pmatrix}$$
$$= \begin{pmatrix} \frac{i\hbar^2}{2} & 0 \\ 0 & -\frac{i\hbar^2}{2} \end{pmatrix}$$
$$= i\hbar \hat{S}_3$$
$$= i\hbar \underbrace{\epsilon_{12k}}_{=0 \text{ except for } k=3} \hat{S}_k$$
$$= i\hbar \epsilon_{123} \hat{S}_3.$$

Here ϵ_{ijk} denotes the totally antisymmetric **Levi-Civita symbol**.

[74] This is what the angular momentum commutation relation (Eq. (3.52)) and the canonical commutation relation (Eq. (3.38)) tell us.

which quantities we use to describe our systems and this is what the next section is about.

3.11 Quantum Numbers

In Classical Mechanics, we describe the angular momentum of an object simply by a vector $\vec{L} = (L_x, L_y, L_z)^T$. Similarly, we usually specify the location *and* the momentum of objects to describe them. In Quantum Mechanics we can't know L_x, L_y and L_z at the same time precisely. At most, we can know *one* of them exactly. And this is really what we do in Quantum Mechanics. We pick just one component of the angular momentum and use it to specify the state of our system. Conventionally, we choose L_z. There is no physical reason behind this since it is completely our choice what we call the z-axis and what the x-axis or y-axis.

However, it turns out that there is a bit more we can know about the angular momentum of a given system: We can also measure the total angular momentum in addition to one component! Classically, the length of the angular momentum vector (squared) is given by $\vec{L}^2 = L_x^2 + L_y^2 + L_z^2$. This quantity tells us the total magnitude of the angular momentum. The corresponding quantum operator is defined analogously

$$\hat{L}^2 = \hat{L}_x^2 + \hat{L}_y^2 + \hat{L}_z^2. \tag{3.58}$$

This quantity is important because we can measure it without affecting the angular momentum components. Again, this property is encoded in a commutation relation: $[\hat{L}^2, \hat{L}_i] = 0$. In words, this means that we can know $\hat{L}^2$ and one $\hat{L}_i$ at the same time with arbitrary precision[75].

[75] Mathematically, we say that $\hat{L}^2$ commutes with $\hat{L}_x$, $\hat{L}_y$ and $\hat{L}_z$.

So to summarize, we specify the angular momentum of a given quantum system using the z-component of the angular momentum $\hat{L}_z$ and the total angular momentum $\hat{L}^2$. It is conventional to use the label m for the angular momentum in the z-direction:

$$\hat{L}_z |m\rangle = \hbar m |m\rangle$$

and the label l for the total angular momentum:[76]

$$\hat{L}^2 \ket{l} = \hbar^2 l(l+1) \ket{l} .$$

In addition, the energy operator $\hat{H}$ often commutes with $\hat{L}_z$ and $\hat{L}^2$, so we can use it as a third label. The conventional label for the energy eigenvalues is n[77].

So in summary: We often label our states using the three labels m, l and n: $\ket{n, m, l}$. Labels like this are known as **quantum numbers**. It is somewhat an art to pick the right quantum numbers for a given problem. However, the basic task is always that we have to find operators that commute with each other $[\hat{A}, \hat{B}] = 0$ since we can only measure operators that commute with each other at the same time precisely.

The whole business becomes even more complicated if we are dealing with a particle with spin. Then, we have in addition to the labels above an additional label for the spin in the z-direction and one for the total spin[78]. In addition, for some systems, it is important how the spin and the orbital angular momentum add up to the total angular momentum.

Similarly, we have to decide between position and momentum labels since $[\hat{p}, \hat{x}] \neq 0$. Here it depends on the problem and experiment. Sometimes we pick the momentum and sometimes the location. However, we can never use both at the same time as labels for our state[79].

Next, we turn to an important question that is still unanswered: how is Quantum Mechanics connected to Classical Mechanics?

The thing is that both yield excellent results which agree with what we observe in nature, however for different kinds of systems. Quantum Mechanics describes how elementary particles behave while Classical Mechanics describes how macroscopic objects behave. But a macroscopic object like a ball also consists

[76] Don't let yourself get confused why we have $l(l+1)$ here instead of simply l. There is a long story behind this, but it's mainly just a convention.

[77] For example, for the Hydrogen atom we have $E_n = -\frac{1}{2} m_e \left(\frac{e^2}{4\pi \varepsilon_0 \hbar} \right)^2 \frac{1}{n^2}$ or for a particle in a box we have $E_n = \frac{n^2 \pi^2 \hbar^2}{D^2 2m}$.

[78] We have these labels additionally since the spin operators commute with the angular momentum operators $[\hat{S}_i, \hat{L}_j] = 0$. In other words, we can measure the orbital angular momentum and spin at the same time we arbitrary precision.

[79] Take note, for example, that the wave functions we discussed in Section 3.3 are only functions of x and t. However, we can switch the labels through a Fourier transform. The result of such a transformation is a $\Psi(p, t)$, i.e., a wave function that only depends on the momentum. For more on this, see Appendix B.

of (many, many) elementary particles. So it must be possible to derive the laws of Classical Mechanics using the laws of Quantum Mechanics by averaging over many many elementary particle systems. This is what the next section is about.

4
The Classical Limit

As already indicated at the end of the last chapter there must be a connection between Quantum Mechanics and Classical Mechanics since classical objects consist of many many quantum objects (e.g., electrons).

We already know the most important concept that allows us to average a large number of elementary particles: the expectation value. And this is exactly the idea we need. The following statement for the momentum expectation value is exactly Newton's second law of Classical Mechanics[1]

$$\frac{d}{dt} \langle \Psi | \hat{p} | \Psi \rangle = - \langle \Psi | \partial_x V(\hat{x}) | \Psi \rangle . \quad (4.1)$$

This equation is known as **Ehrenfest's theorem**. We will discuss below why it is true. In words, this theorem tells us that the quantum mechanical expectation values obey Newton's classical equations of motion[2].

Now, as promised, we derive Eq. (4.1).

We start with the expectation value for a general operator $\hat{O}$

---

[1] Recall that the definition of a force is $F(x) \equiv -\partial_x V(x)$ and Newton's second law is $F = \partial_t p = \partial_t mv = ma$, at least if the mass m is constant. (A system where this is not the case is, for example, a rocket which progressively gets lighter through the burning of fuel.)

[2] Strictly speaking, this is not always true. It is only correct for simple systems where for the potential we have $\langle \Psi | \partial_x V(\hat{x}) | \Psi \rangle = \partial_x V(\langle \Psi | \hat{x} | \Psi \rangle)$. This equation is true for potentials that are of most second order in x. If the potential contains higher order terms like x^3 etc. we can see immediately that it is no longer correct:

$$V = a + bx + cx^2 + dx^3$$
$$\therefore \quad \partial_x V = b + 2cx + 3dx^2$$
$$\therefore \quad \langle \partial_x V \rangle = \langle b + 2cx + 3dx^2 \rangle$$
$$\neq b + 2c\langle x \rangle + 3d\langle x \rangle^2 .$$

We already learned in Section 3.2 that it makes a huge difference whether we square an expectation value: $\langle x \rangle^2$ or if we calculate the expectation value of a squared quantity: $\langle x^2 \rangle \neq \langle x \rangle^2$! For example, for $\langle x^2 \rangle$ we only sum over positive values since we squared them, while for $\langle x \rangle^2$ there are also negative values in the sum and we only square the result of the sum. In contrast, if the potential is of second order in x we have no such problem since $\partial_x V$ no term proportional to x^2 appears.

For systems with higher order potentials our Eq. (4.1) is only valid for small excitations in a region of the potential where we can approximate it by a polynomial which is at most second order in x.

(Eq. (3.27))
$$\langle\Psi|\hat{O}|\Psi\rangle = \int d^3x \Psi^\star \hat{O}\Psi,$$

the Schrödinger equation (Eq. (3.43))
$$\frac{d}{dt}\Psi = \frac{1}{i\hbar}H\Psi \qquad (4.2)$$

and the complex conjugated Schrödinger equation

$$(\text{Eq. 4.2})^\dagger \leftrightarrow \frac{d}{dt}\Psi^\dagger = -\frac{1}{i\hbar}\Psi^\dagger \overbrace{H^\dagger}^{H^\dagger=H}$$
$$= -\frac{1}{i\hbar}\Psi^\dagger H. \qquad (4.3)$$

Taking the time derivative of the expectation value then yields

$$\frac{d}{dt}\langle\hat{O}\rangle = \frac{d}{dt}\int d^3x\, \Psi^\dagger \hat{O}\Psi$$
$$\underbrace{=}_{\text{product rule}} \int d^3x \left(\left(\frac{d}{dt}\Psi^\dagger\right)\hat{O}\Psi + \Psi^\dagger\left(\frac{d}{dt}\hat{O}\right)\Psi + \Psi^\dagger\hat{O}\left(\frac{d}{dt}\Psi\right)\right).$$

Next, we use $\frac{d}{dt}\hat{O} = 0$, which is correct for most operators. For example, for $\hat{O} = \hat{p} = -i\hbar\vec{\nabla} \neq \hat{O}(t)$. In addition, we use the Schrödinger equation to rewrite the time derivatives of the wave function and its complex conjugate. This yields

$$\frac{d}{dt}\langle\hat{O}\rangle = \int d^3x \left(\left(\frac{d}{dt}\Psi^\dagger\right)\hat{O}\Psi + \underbrace{\Psi^\dagger\left(\frac{d}{dt}\hat{O}\right)\Psi}_{=0} + \Psi^\dagger\hat{O}\left(\frac{d}{dt}\Psi\right)\right)$$
$$\underbrace{=}_{\text{Eq. 4.2 and Eq. 4.3}} \int d^3x \left(\left(-\frac{1}{i\hbar}\Psi^\dagger H\right)\hat{O}\Psi + \Psi^\dagger\hat{O}\left(\frac{1}{i\hbar}H\Psi\right)\right)$$
$$= \frac{1}{i\hbar}\int d^3x(-\Psi^\dagger H\hat{O}\Psi + \Psi^\dagger \hat{O}H\Psi)$$
$$= \frac{1}{i\hbar}\int d^3x\, \Psi^\dagger[\hat{O}, H]\Psi$$
$$= \frac{1}{i\hbar}\langle[\hat{O}, H]\rangle, \qquad (4.4)$$

Where we used in the last step that in the second to last line we have exactly the formula for the expectation value (Eq. (3.27)) for the operator $[\hat{O}, H]$.[3] Now we evaluate this equation specif-

---

[3] Recall, that in general, the commutator of two operators yields again a generator. For example, the commutator of the position and location operator yields the trivial identity operator: $[\hat{p}_i, \hat{x}_j] = -i\hbar\delta_{ij}$ (Eq. (3.38)), or the commutator of the spin operators yields another spin operator: $[\hat{S}_i, \hat{S}_j] = i\hbar\epsilon_{ijk}\hat{S}_k$ Eq. (3.57).

ically for the momentum operator $\hat{O} = \hat{p}$ and, in addition, use the explicit form of the Hamiltonian operator $H = \frac{\hat{p}^2}{2m} + V$:

$$\begin{aligned}
\frac{d}{dt}\langle \hat{p} \rangle &= \frac{1}{i\hbar}\langle [\hat{p}, H] \rangle \\
&= \frac{1}{i\hbar}\langle [\hat{p}, \frac{\hat{p}^2}{2m} + V] \rangle \\
&= \frac{1}{i\hbar}\langle \underbrace{[\hat{p}, \frac{\hat{p}^2}{2m}]}_{=0} + [\hat{p}, V] \rangle \\
&= \frac{1}{i\hbar}\langle [\hat{p}, V] \rangle \\
&= \frac{1}{i\hbar}\int d^3x \Psi^\dagger [\hat{p}, V] \Psi \\
&= \frac{1}{i\hbar}\int d^3x \Psi^\dagger \hat{p} V \Psi - \frac{1}{i\hbar}\int d^3x \Psi^\dagger V \hat{p} \Psi \\
&= \frac{1}{i\hbar}\int d^3x \Psi^\dagger (-i\hbar \boldsymbol{\nabla}) V \Psi - \frac{1}{i\hbar}\int d^3x \Psi^\dagger V (-i\hbar \boldsymbol{\nabla}) \Psi \quad \text{(using Eq. 3.36)} \\
&= -\int d^3x \Psi^\dagger (\nabla V) \Psi - \int d^3x \Psi^\dagger V \nabla \Psi + \int d^3x \Psi^\dagger V \nabla \Psi \quad \text{(product rule)} \\
&= -\int d^3x \Psi^\dagger (\nabla V) \Psi \\
&= \langle -\nabla V \rangle = \langle F \rangle. \quad (4.5)
\end{aligned}$$

This is exactly the equation stated at the beginning of this section.

Now, it's time to summarize what we have learned so far before we move on and discuss concrete quantum systems.

5
Summary

We started with a short discussion of the infamous double slit experiment. The lesson we learned is that we need waves to describe particles.

One immediate consequence of this observation is that for some quantities only a discrete set of values is possible[1]. We say they are quantized.

[1] One example we discussed was the number of wave crests you can produce in a rope.

A second consequence of our need for waves to describe particles is that we get a fundamental quantum uncertainty. For example, each time we measure the momentum we change the location and vice versa. Therefore, it's impossible to know the momentum and location of a particle at the same time with arbitrary precision.

After this preliminary discussion of the most important features of Quantum Mechanics, we started to think about how we can describe them within a physical theory.

The basic idea was to introduce an abstract object $|\Psi\rangle$ which describes the system in question. In addition, we introduced quantum operators that we can use to extract information about the system. For example, if we want to know the momentum of a system, we use the momentum operator $\hat{p}$:

$$\hat{p}\ket{\Psi_1} = p_1 \ket{\Psi_1} .$$

However, for quantum systems, we often do not get such a simple answer. Instead, if we measure the momentum of equally prepared systems, we possibly end up with different results. Each possible result arises with a certain probability. In our quantum framework we describe such a situation like this

$$\ket{\Psi} = a\ket{\Psi_1} + b\ket{\Psi_2} + \ldots ,$$

Here $\ket{\Psi_1}$ is a state with momentum p_1, and $\ket{\Psi_2}$ the state with momentum p_2. The coefficients a and b are directly related to the probability to measure a or b. For example, $|a|^2$ is the probability to measure the value p_1.

Afterwards, we took a short side trip into the world of statistics. We discussed two of the most important notions that we need all the time in Quantum Mechanics: the expectation value and the standard deviation.

We then discussed how we can calculate the expectation value explicitly for quantum systems. The basic idea is that we "sandwich" the corresponding operator between a ket $\ket{\Psi}$ and a bra $\bra{\Psi}$. For example, the momentum expectation value reads[2]

$$\bra{\Psi}\hat{p}\ket{\Psi} .$$

In addition, we learned how we can calculate the probability to measure one *specific* value. All we have to do is multiply the ket that describes our state with the bra that describes the state with exactly this value. For example, the probability to measure p_1 is
[3]

$$|\braket{\Psi_1|\Psi}|^2 .$$

Then we talked about wave functions. A wave function is simply the set of coefficients if we expand our state $\ket{\Psi}$ in one specific basis, namely the position basis

$$\ket{\Psi} = \int dx \Psi(x) \ket{x} . \tag{5.1}$$

[2] Given a ket, we can immediately calculate the corresponding bra. A bra together with a ket denotes the scalar product between two abstract vectors. This is analogous to $\braket{\vec{v}_1|\vec{v}_2} \equiv \vec{v}_1 \cdot \vec{v}_2$. Here $\bra{\vec{v}_1} \equiv \vec{v}_1^T$ ("row times column"). But in Quantum Mechanics we deal with complex vectors and therefore have $\bra{\Psi}_1 \equiv \ket{\Psi_1}^\dagger = \ket{\Psi_1}^{*T}$.

[3] Recall that this is completely analogous to how we can calculate how much a vector spreads out in the, say, z-direction. All we have to do is multiply it by the $\vec{e}_z$ basis vector.

SUMMARY

Afterwards, we thought about the crucial question: what do quantum operators explicitly look like?

As a first step, we recalled the basic message of Noether's theorem: symmetries lead to conserved quantities. For example, symmetry under spatial translations leads to conservation of momentum. Then we took another short side trip and talked about symmetries in general. The basic lesson here was that the basic mathematical object that we need to describe continuous symmetries are the so-called generators. By acting with these generators on an object, we can achieve an infinitesimal transformation. By repeating this tiny transformation many times, we can generate any transformation that we like. We then connected these two puzzle pieces and proposed that the quantum operators are precisely the generators we just talked about. For example, the quantum momentum operator is simply the generator of spatial translations: $\hat{p} \equiv -i\hbar \partial_x$. The second most important example is the quantum energy operator $\hat{E} \equiv i\hbar \partial_t$. We concluded that this is the correct explicit form of the operator since symmetry under temporal translation leads to conservation of energy.[4]

[4] A temporal translation means that we shift the time when our experiment happens: $t \to t + a$.

These ideas allowed us to derive the two most important quantum equations:

▷ The canonical commutation relation: $[\hat{p}_i, \hat{x}_j] = -i\hbar \delta_{ij}$. This equation tells us that it really makes a difference whether we first measure the momentum of our system or first the location[5].

[5] Reminder: $[\hat{A}, \hat{B}] = \hat{A}\hat{B} - \hat{B}\hat{A}$.

▷ The Schrödinger equation: $i\hbar \partial_t |\Psi\rangle = -\frac{\hbar^2 \partial_i^2}{2m} |\Psi\rangle + V(\hat{x}) |\Psi\rangle$. This equation tells us how quantum systems evolve in time. In addition, it allowed us to understand how waves play a role in our framework. The crucial observation here was that solutions of the Schrödinger equation behave like waves.

Finally, we talked about the third most important quantum operator: angular momentum. Noether's theorem tells us that angular momentum is conserved if the system does not change under rotations. Following the same logic as above, we use the generator of rotations as our quantum angular momentum operator. However, there is a subtlety since a rotation possibly has two effects. If we describe our system with an object that has more than one component we not only need to modify the argument of our function $\Psi(x) \to \Psi(Rx)$, where Rx denotes the rotated coordinates. Instead, additionally we need to take into account that components possibly get mixed:

$$\begin{pmatrix} \Psi_1(x,t) \\ \Psi_2(x,t) \end{pmatrix} \to \begin{pmatrix} \Psi_2(x,t) \\ \Psi_1(x,t) \end{pmatrix}.$$

Therefore, we have two generators of rotations[6].

[6] The complete generator of rotations is given by a combination of these two generators.

▷ One generator is given by a differential operator $\hat{L} \equiv \vec{x} \times (-i\vec{\partial})$ and is responsible for the transformation $\Psi(x) \to \Psi(Rx)$.

▷ The second generator is given by a matrix, for example[7], $\frac{1}{2}\sigma_i$ and is responsible for the mixing of the components.

[7] Recall that σ_i denotes the Pauli matrices. This second generator only looks like this if our system has spin 1/2 and we can therefore use two-component objects to describe it. For spin 1 systems, for example, we need a four-component object and (4×4) matrices for the generators.

Consequently, we therefore also have two different quantum operators for angular momentum. The first one describes orbital angular momentum, which describes how one objects revolves around another. The second one describes some kind of internal angular momentum known as spin. The unusual thing about this new kind of angular momentum is that only a discrete set of values are possible since the operator is given by a matrix. In other words, spin is quantized.

Now it's time to see how all this works in practice. In the following sections, we will talk about the most important quantum systems and how we can describe them using the tools that we just talked about.

Part II
Essential Quantum Systems and Tools

"A mathematician may say anything he pleases, but a physicist must be at least partially sane."

Josiah Willard Gibbs

PS: You can discuss the content of Part II with other readers, find exercises to check your understanding and give feedback at www.nononsensebooks.com/qm/part2.

In the following chapters, we will discuss several quantum systems in detail. However, we will only discuss a quite small number of system since these are enough to understand almost all crucial aspects of Quantum Mechanics. The quantum systems we will discuss in this part are

▷ A particle in a one-dimensional box with *infinite* walls. This example demonstrates nicely how and why energy is quantized in some quantum systems.

▷ A particle in a one-dimensional box with *finite* walls. This example is similar to the previous one, but additionally, we can learn here how quantum particles can tunnel into regions that are classically forbidden.

▷ The Hydrogen atom. This system is, in some sense, a three-dimensional "box" and the most important new aspect here is that angular momentum plays a role.

▷ A particle that scatters *off* a box. While the previous three examples are all about bound states, this example illustrates how we can describe scattering states.

While there are infinitely many different variations of these problems (differently shaped boxes), the systems in the list above are absolutely sufficient to grasp the fundamental features of Quantum Mechanics. Afterwards, we will talk about everyone's darling:

▷ The harmonic oscillator. This example is super important since if we only care about small movements, many potentials look *exactly* like the potential of the harmonic oscillator. Also, there are two different ways to treat the oscillator. The first method is completely analogous to what we did in the other examples. However, there is a second method, which is a lot more clever and essential if you want to understand Quantum Field Theory[8]. In fact, oversimplifying a bit, we can say that quantum fields are just a collection of lots of harmonic oscillators.[9]

[8] Quantum Field Theory is what we end up with if we combine the fundamental lessons of Quantum Mechanics with the fundamental lessons of Einstein's theory of Special Relativity.

[9] Think: like a mattress. At each point in space we have a spring (= a harmonic oscillator) and all these springs are connected. It is indeed possible to derive the correct Lagrangian for a scalar field by starting with a discrete set of coupled harmonic oscillators (= a mattress) and then taking the continuum limit.

Before we start, let's discuss a few general things that are extremely useful to keep in mind.

6

Tricks and Ideas We Need All the Time

What we do in Quantum Mechanics is usually always the same:

We solve the Schrödinger equation

$$i\hbar\partial_t \Psi(x) = \left(-\frac{\hbar^2 \partial_x^2}{2m} + V(x)\right)\Psi(x)$$

for some given potential $V(x)$ and boundary conditions. Boundary conditions are always an essential physical input. The boundary conditions determine, for example, if a particle moves towards our box from the left or the right-hand side, or if a given particle is confined inside the potential or not.

The solutions then tell us which energy values are possible and, for example, where it is most likely that we find our particle.

Now, one of the things we should talk about before we start with concrete examples is a clever trick that makes the task of solving the Schrödinger equation much simpler.

6.1 Let's Separate Time and Space

For any system where the potential does not depend on time, we can split the Schrödinger equation into two simpler equations and solve them separately. This trick is known as **separation of the variables** and works as follows.

First, we split our wave function into two parts:

$$\Psi(x,t) \equiv T(t)\psi(x). \tag{6.1}$$

The first part $T(t)$ describes how the wave function changes over time, while the second part $\psi(x)$ how the wave function depends on the location. We now put this ansatz into the Schrödinger equation:[1]

$$i\hbar \Psi(x,t) = H\Psi(x,t)$$
$$\therefore \quad i\hbar \partial_t T(t)\psi(x) = HT(t)\psi(x) \quad \text{(using Eq. 6.1)}$$
$$\therefore \quad i\hbar \frac{\partial_t T(t)}{T(t)} = \frac{H\psi(x)}{\psi(x)}.$$

[1] Reminder: For a particle in a potential $V(x)$, the Schrödinger equation reads

$$i\hbar \frac{\partial \Psi}{\partial t} = -\frac{\hbar^2}{2m}\frac{\partial^2 \Psi}{\partial x^2} + V(x)\Psi.$$

The operator H is the energy operator and it is directly related to the classical energy = kinetic energy plus potential energy.

In the last step we used that H only contains the spatial derivative ∂_x^2 and $T(t)$ only depends on t and not on x. In addition, we used that the left-hand side only contains the derivative ∂_t while $\psi(x)$ only depends on x.

Now what we are left with here is on the left-hand side something that only depends on t, and on the right-hand side something that only depends on x. Still, both sides must be equal. This is only possible if both sides are constant. If, for example, the left-hand side is non-constant this would mean that we could change its value by varying t. Since the right-hand side does not depend on t at all, there is then no way that both sides are equal. We call the constant both sides of the equation are equal to E (= the energy). Then we are left with two equations, as promised:

$$i\hbar \frac{\partial_t T(t)}{T(t)} \equiv E \equiv \frac{H\psi(x)}{\psi(x)}$$
$$\therefore \quad i\hbar \frac{\partial_t T(t)}{T(t)} \equiv E \quad \text{and} \quad \frac{H\psi(x)}{\psi(x)} \equiv E,$$

or written a bit differently

$$i\hbar \partial_t T(t) = E T(t) \tag{6.2}$$
$$H\psi(x) = E\psi(x). \tag{6.3}$$

The first equation is easy to solve and does not depend on the specific problem at all. All information about the system are encoded in H. So the first lesson here is that, as long as V (and therefore also H) does not depend on t,[2] the explicit time-dependence of the wave function is given by

$$\boxed{T(t) = \exp\left(-\frac{iEt}{\hbar}\right).} \tag{6.4}$$

[2] Remember that this was the crucial restriction we mentioned at the beginning. The whole trick we used here only works if H does not depend on t. Otherwise, the ansatz $\Psi(x,t) \equiv T(t)\psi(x)$ does not help us.

since

$$i\hbar \partial_t T(t) = E T(t) \quad \text{(this is Eq. 6.2)}$$
$$\therefore \quad i\hbar \partial_t \exp\left(-\frac{iEt}{\hbar}\right) = E \exp\left(-\frac{iEt}{\hbar}\right) \quad \text{(using Eq. 6.4)}$$
$$\therefore \quad i\hbar \left(\frac{-iE}{\hbar}\right) \exp\left(-\frac{iEt}{\hbar}\right) = E \exp\left(-\frac{iEt}{\hbar}\right)$$
$$\therefore \quad E \exp\left(-\frac{iEt}{\hbar}\right) = E \exp\left(-\frac{iEt}{\hbar}\right) \checkmark$$

The second lesson is that all specific information about the system are encoded in the solutions of the second equation which only depends on x:[3]

$$\boxed{H\psi = E\psi} \tag{6.5}$$

[3] Take note that this equation is really of the same type as the equations we considered all the time (an eigenvalue equation). The energy operator H acts on Ψ and what we get back is the energy E.

This equation is known as the **stationary Schrödinger equation** or **time-independent Schrödinger equation**. After we have solved it for a specific problem (i.e. a specific H), all that is left to do is to remember that the full solution reads

$$\Psi(x,t) = \psi(x) T(t) = \psi(x) \exp\left(-\frac{iEt}{\hbar}\right). \tag{6.6}$$

A solution of the stationary Schrödinger equation $\psi(x)$ is known as a **stationary solution** or a **stationary state**.

It is clear that this kind of problem can become arbitrarily complicated for complicated potentials $V(x)$. However, there is often not much new physics to be learned in such complicated problems and we will stick to the simplest potentials.[4]

[4] As already mentioned in Section 3.8 the general mathematical theory that deals with this kind of problems is known as spectral theory.

The next thing we need to talk about before we start with concrete quantum systems is an important restriction that tells us which solutions of the Schrödinger equation make sense physically.

6.2 Why Quantum Waves are Smooth

While mathematically lots of solutions are possible, in physics only a few of them make sense. An incredibly important restriction on physically allowed solutions is that our wave function has to be smooth everywhere. Formulated differently: No jumps in the wave function are allowed. This restriction follows since the momentum operator is proportional to the derivative operator ∂_x. Therefore, if there are any jumps in the wave function, the derivative and thus also the momentum at these points is infinite. An infinite momentum is not physical and therefore no jumps are allowed. Therefore, we can conclude that the wave function has to be smooth everywhere.

Analogously, the first derivative of the wave function $\partial_x \Psi_x$ must be smooth, too. This restriction follows since the formula for the kinetic energy is $\hat{T} = \hat{p}^2/2m = -\hbar^2 \partial_x^2/2m$. A jump in the first derivative $\partial_x \Psi_x$ would mean that the kinetic energy is infinite at these points[5].

[5] There is an important exception to this rule. If we deal with a system with a potential that is infinite at at least one point, the first derivative is allowed to jump at this point. The reason for this is that if we introduce something non-physical like an infinite potential energy, an infinite kinetic energy becomes possible. Such models are nothing we can observe in nature but only useful as toy models that help us to understand important aspects in a simplified setup.

Now we are *almost* ready to dive in and calculate explicitly what happens in quantum systems. There is only one more thing we need to talk about. The following classification is immensely helpful to make sense of the solutions that we calculate in the following sections.

6.3 Classification of Solutions

We already talked about one possible solution of the Schrödinger equation in Section 3.7. There we investigated the Schrödinger equation

$$i\hbar \partial_t \Psi(x,t) = -\frac{\hbar^2 \partial_x^2}{2m} \Psi(x,t) + V(\hat{x})\Psi(x,t) \qquad (6.7)$$

when there is no potential $V(\hat{x}) = 0$ and found that in this case solutions look like this

$$\Psi(x,t) = e^{-i(Et-px)/\hbar} .$$

We call this kind of solution a plane wave since it describes a wave that spreads out all over space with constant amplitude.[6]

Now, if there is a potential we need, of course, different solutions. However, the solutions not only depend on the potential $V(\hat{x})$ but also on the energy of the particle E. To understand this, let's consider for simplicity the stationary Schrödinger equation (Eq.6.5):

$$E\psi(x) = -\frac{\hbar^2 \partial_x^2}{2m}\psi(x) + V(\hat{x})\psi(x) . \qquad (6.8)$$

The free solution (i.e. the solution for $V(\hat{x}) = 0$) of the stationary Schrödinger equation follows directly from our result above

$$\psi(x) = e^{(ipx)/\hbar} ,$$

where $p = \sqrt{2mE}$.

Now, what happens if the potential is nonzero?[7]

[6] As a further reminder: such plane waves do not describe physical particles since they spread out all over space. To describe a real particle we need to use superpositions of plane waves. A suitable superposition of plane waves yields a wave packet which we can use to describe a particle.

[7] Take note that, in general, the potential has not the same value everywhere. Hence the solutions we discuss below can be valid for different regions of the same system. For example, imagine a box where the potential is zero inside and non-zero outside the box.

▷ If the energy of the particle is *larger* than the potential $E > V$, we get again an oscillating solution but with a different frequency:[8]

$$\psi(x) = e^{(i\tilde{p}x)/\hbar},$$

where $\tilde{p} = \sqrt{2m(E-V)}$.

[8] We will check this explicitly in a later section. Here, we are only intersted in the general strucutre.

▷ If the energy of the particle is *smaller* than the potential $E < V$, our solution is, in principle, also

$$\psi(x) = e^{(i\tilde{p}x)/\hbar}.$$

However, we need to be careful here since upon closer inspection this solution is qualitatively completely different. This follows since now the term under the square root in $\tilde{p} = \sqrt{2m(E-V)}$ is negative. This means that for $E < V$, we can factor out a -1 from under the square root:

$$\tilde{p} = \sqrt{2m(E-V)} = \sqrt{(-1) \times 2m(V-E)}$$
$$= \sqrt{-1}\sqrt{2m(V-E)} = i\sqrt{2m(V-E)}.$$

We then define $\rho \equiv \sqrt{2m(V-E)}$ and therefore now have

$$\tilde{p} = i\rho. \tag{6.9}$$

Our solution then reads

$$\psi(x) = e^{(i\tilde{p}x)/\hbar} = e^{(ii\rho x)/\hbar} = e^{(-\rho x)/\hbar}.$$

The crucial point here is that if $E < V$, the argument of the exponential function is no longer imaginary. While an exponential function with imaginary argument describes an **oscillating wave**, an exponential function with real argument describes exponential growth or **exponential decay**. In physically terms this usually means that our wave function quickly becomes tiny in regions where $E < V$.[9]

[9] This will make a lot more sense as soon as we discuss concrete examples

▷ A third possibility is that the potential is infinite: $V(\hat{x}) = \infty$. In this case, the only possible solution of the stationary Schrödinger equation is $\psi(x) = 0$. In physical terms this means that the probability to find a particle in a region where the potential is infinitely high is zero.

With this in mind, we can understand the following general classification of solutions of the Schrödinger equation.

▷ If the energy of the state is smaller than the potential in the system at infinity: $E < V(\infty)$ and $E < V(-\infty)$, we are dealing with a **bound state**. Such states describe a particle which is stuck inside some kind of box. This interpretation comes about since, as discussed above, solutions for the Schrödinger equation with $E < V$ describe either an exponential decay or exponential growth. In the former case, we get a vanishing wave function at $x = \infty$ and in the latter case an infinite value for the wave function. An infinite value of the wave function makes no sense. Therefore, states with $E < V(\infty)$ and $E < V(-\infty)$ cannot "come from infinity", but describe particles confined within some fixed region. We label bound states using a *discrete* index n. We use discrete labels since the energy levels within a box are quantized and therefore there is a discrete set of possible states. A general solution is a *sum* of such solutions.

▷ If the energy of the state is larger than the potential at infinity: $E > V(\infty)$ or $E > V(-\infty)$, we are dealing with a **scattering state**. We can understand this interpretation by recalling that for $E > V$ we get oscillating solutions. Therefore, such states describe waves that come "from infinity" and then slow down under the influence of some potential or even get reflected. We label scattering states using a *continuous* index k. A general solution is an *integral* $\int dk$ over such solutions[10].

Now with this in mind, we are finally ready to discuss the most important examples explicitly. But before we dive in one more comment: Don't worry if not every step is clear at first glance. Just keep reading and things will make more and more sense if you see them explained multiple times for different problems. There will be quite some repetition in the following sections. This is on purpose to help you internalize the basic methods. So the goal is not that you understand everything immediately, but instead that you slowly get used to how people talk about quantum systems.

[10] An important subtlety is that since such wave functions do not vanish at infinity, they are non-normalizable. This tells us immediately that a pure scattering wave function is not in their own right physically realizable. Instead, only suitable linear combinations (wave packets) are physically realizable. However, often it is sufficient to investigate the basic building blocks (= plane waves) that our wave packets consist of. This is good news since dealing with wave packets is a messy business and seldom possible without a computer.

7
Quantum Mechanics in a Box

7.1 The Infinite Box

A particle confined in a box, here 1-dimensional, with infinitely high potential walls is one of the most famous quantum systems. Inside the box the potential is zero, outside it's infinite:

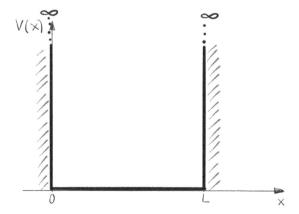

[1] Here we will not use the stationary Schrödinger equation (Eq. (3.41)) since it is instructive to see at least one example worked out with the explicit time dependence. However, in the following sections, we will always start directly with the stationary Schrödinger equation. This is possible because the time dependence is always the same for these systems since the potentials do not change over time.

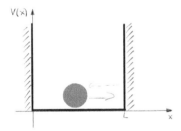

Figure 7.1: A classical ball in a box would simply bounce between the walls.

[2] We talked already about the free particle solution in Section 3.3.

[3] Recall: If there are any jumps in the wave function, the momentum of the particle $\hat{p}_x \Psi = -i\partial_x \Psi$ is infinite because the derivative at the jumping point would be infinite.

The potential is defined piece-wise

$$V = \begin{cases} 0, & 0 < x < L \\ \infty, & \text{otherwise} \end{cases} \quad (7.1)$$

and therefore, we have to solve the one-dimensional Schrödinger equation[1]

$$i\hbar \partial_t \Psi(\vec{x},t) = -\frac{\hbar^2 \partial_x^2}{2m}\Psi(x,t) + V(x)\Psi(x,t)$$

piece-wise.

▷ Inside the box, the solution is equal to the free particle solution, because $V = 0$ for $0 < x < L$

▷ Outside the box, since $V = \infty$, the only possible, physical solution is $\Psi(x,t) = 0$.

We can rewrite the general free particle solution[2]

$$\Psi(x,t) = A e^{-i(Et-px)/\hbar} + B e^{-i(Et+px)/\hbar}$$
$$= \big(C \sin((px)/\hbar) + D \cos((px)/\hbar)\big) e^{-iEt/\hbar},$$

using the non-relativistic energy-momentum relation

$$E = \frac{p^2}{2m} \quad \to \quad p = \sqrt{2mE} \quad (7.2)$$

as

$$\Psi(x,t) = \big(C \sin((px)/\hbar) + D \cos((px)/\hbar)\big) e^{-iEt/\hbar}$$
$$= \Big(C \sin\big(\frac{\sqrt{2mE}}{\hbar} x\big) + D \cos\big(\frac{\sqrt{2mE}}{\hbar} x\big)\Big) e^{-iEt/\hbar} \quad \text{(using Eq. 7.2)}$$

Next, we use that the wave function must be a continuous function[3]. Therefore, we have the boundary conditions

$$\Psi(0,t) = \Psi(L,t) \stackrel{!}{=} 0. \quad (7.3)$$

First of all, because $\cos(0) = 1$ we can conclude $D \stackrel{!}{=} 0$. Furthermore, we see that these two conditions imply

$$\sqrt{\frac{2mE}{\hbar^2}} \stackrel{!}{=} \frac{n\pi}{L}, \quad (7.4)$$

with arbitrary integer $n = 1, 2, 3, \ldots$, because only then

$$\Psi_n(x,t) = \left(C\sin(\frac{\sqrt{2mE}}{\hbar}x) + \underbrace{D}_{=0}\cos(\frac{\sqrt{2mE}}{\hbar}x)\right)e^{-iEt/\hbar}$$
$$= C\sin\left(\frac{n\pi}{L}x\right)e^{-iE_n t/\hbar} \qquad (7.5)$$

both boundary conditions are satisfied:

$$\Psi_n(0,t) \stackrel{!}{=} 0$$
$$\text{check:} \quad \Psi_n(0,t) = C\underbrace{\sin\left(\frac{n\pi}{L}0\right)}_{=0}e^{-iE_n t/\hbar}$$
$$= 0 \checkmark$$

and

$$\Psi_n(L,t) \stackrel{!}{=} 0$$
$$\text{check:} \quad \Psi_n(L,t) = C\sin\left(\frac{n\pi}{\cancel{L}}\cancel{L}\right)e^{-iE_n t/\hbar}$$
$$= C\underbrace{\sin(n\pi)}_{=0}e^{-iE_n t/\hbar}$$
$$= 0 \checkmark$$

We can rewrite Eq. (7.4) to find a condition for the energy E

$$E_n \stackrel{!}{=} \frac{n^2\pi^2\hbar^2}{L^2 2m}. \qquad (7.6)$$

The possible energies are therefore quantized, which means that the corresponding allowed energy levels are integer multiples of the constant: $\frac{\pi^2\hbar^2}{L^2 2m}$.

Take note that we have a solution for each n and linear combinations of the form

$$\Psi(x,t) = A\Psi_1(x,t) + B\Psi_2(x,t) + \ldots$$

are solutions, too.

We can calculate the normalization constant C by using that the probability P for finding the particle anywhere inside the box

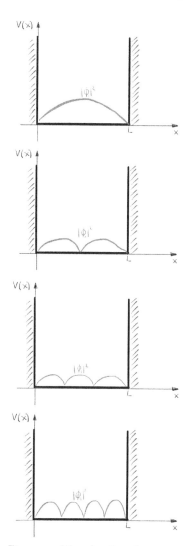

Figure 7.2: Wave functions in the infinite box. Take note that our wave function is in general a complex function and plotting a complex function is difficult. Therefore, here the absolute square of the wave function is shown. This absolute square indicates the probability to find the particle at the given locations.

must be 100% = 1 and the probability outside is zero, because there we have $\Psi = 0$. Therefore

$$1 = \int_0^L dx \Phi_n^\star(x,t) \Phi_n(x,t)$$

$$= \int_0^L dx C^2 \sin(\frac{n\pi}{L}x) e^{+iEt} \sin(\frac{n\pi}{L}x) e^{-iEt} \quad \text{(using Eq. 7.5)}$$

$$= C^2 \int_0^L dx \sin^2(\frac{n\pi}{L}x)$$

$$= C^2 \left[\frac{x}{2} - \frac{\sin(\frac{2n\pi}{L}x)}{4\frac{n\pi}{L}} \right]_0^L$$

$$= C^2 \left(\frac{L}{2} - \frac{\sin(\frac{2n\pi}{L}L)}{4\frac{n\pi}{L}} \right)$$

$$= C^2 \frac{L}{2}$$

and we can conclude

$$C^2 = \frac{2}{L} \quad (7.7)$$

With all this in mind, we can now calculate, for example, the expectation value of the position operator for various states of the system. Let's assume the particle is in the ground state:

$$\Psi_1(x,t) = \sqrt{\frac{2}{L}} \sin\left(\frac{\pi}{L}x\right) e^{-iE_n t/\hbar} \quad \text{((Eq. (7.5) with } n=1\text{))}. \quad (7.8)$$

The position expectation value is then (Eq. (3.27)):[4]

$$\langle \Psi_1 | \hat{x} | \Psi_1 \rangle = \int_0^L dx \Psi_1^\dagger(x,t) \hat{x} \Psi_1(x,t)$$

$$= \int_0^L dx \left(\sqrt{\frac{2}{L}} \sin\left(\frac{\pi}{L}x\right) e^{-iE_n t/\hbar} \right)^\dagger x$$

$$\times \left(\sqrt{\frac{2}{L}} \sin\left(\frac{\pi}{L}x\right) e^{-iE_n t/\hbar} \right)$$

$$= \frac{2}{L} \underbrace{\int_0^L dx \sin^2\left(\frac{\pi}{L}x\right) x}_{=\frac{L^2}{4}}$$

$$= \frac{L}{2} \quad (7.9)$$

In words this result tells us that the most probable position for the particle in the ground state $\Psi_1(x,t)$ is exactly in the middle.

[4] We integrate from $-\infty$ to ∞. However, the wave function is zero outside the box. Therefore we end up with the nonzero part of the integral, which goes from 0 to L.

Moreover, we can now calculate the standard deviation (Eq. (3.7)) explicitly

$$\Delta x = \sqrt{\langle x^2 \rangle - \langle x \rangle^2}$$
$$= \sqrt{\langle x^2 \rangle - \left(\frac{L}{2}\right)^2} \quad \text{(using Eq. 7.9)}$$

So what we still need here is the expectation value of the squared position operator: $\langle x^2 \rangle$. We can calculate it directly[5]

[5] You can do the integration, for example, using http://wolframalpha.com.

$$\langle \Psi_1 | \hat{x}^2 | \Psi_1 \rangle = \int_0^L dx \, \Psi_1^\dagger(x,t) \hat{x}^2 \Psi_1(x,t)$$
$$= \int_0^L dx \left(\sqrt{\frac{2}{L}} \sin\left(\frac{\pi}{L}x\right) e^{-iE_n t/\hbar} \right)^\dagger x^2 \left(\sqrt{\frac{2}{L}} \sin\left(\frac{\pi}{L}x\right) e^{-iE_n t/\hbar} \right)$$
$$= \frac{2}{L} \underbrace{\int_0^L dx \sin^2\left(\frac{\pi}{L}x\right) x^2}_{\approx 0.14 L^3}$$
$$\approx 0.28 L^2 . \qquad (7.10)$$

With this information at hand, we find

$$\Delta x = \sqrt{\langle x^2 \rangle - \langle x \rangle^2}$$
$$= \sqrt{\langle x^2 \rangle - \left(\frac{L}{2}\right)^2} \quad \text{(using Eq. 7.9)}$$
$$\approx \sqrt{0.28 L^2 - \frac{L^2}{4}} \quad \text{(using Eq. 7.10)}$$
$$= L \sqrt{0.28 - 0.25}$$
$$= 0.17 L .$$

This result tells us that, on average, we will find the particle will around 17% of the total box width away from the middle.

Completely analogous we can calculate the expectation value and standard deviation for all other possible states. For example, for the second lowest state Ψ_2 the expectation value is again exactly in the middle of the box ($\langle \Psi_2 | \hat{x} | \Psi_2 \rangle = \frac{L}{2}$), but the standard deviation is larger: $\Delta x \approx 0.27L$. This tells us that if the particle is in this state, it fluctuations a lot more, which is exactly what we would expect from a state with higher energy ($E_2 > E_1$).

7.2 The Finite Box

The system we discussed in the last section was somewhat oversimplified. There are no infinite boxes in nature. Therefore, it makes sense to have a look at a similar system with finite potential barriers. Concretely, the potential we are now interested in reads

$$V = \begin{cases} 0, & -a < x < a \\ V_0, & \text{otherwise} \end{cases} \quad (7.11)$$

and looks like this

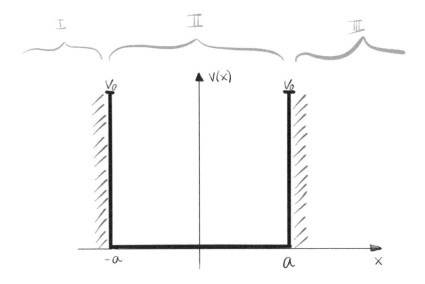

Our only task is to solve the *stationary* Schrödinger equation (Eq. (6.5))

$$\frac{d^2\Psi}{dx^2} = \frac{2m}{\hbar^2}[V(x) - E]\Psi$$

since the potential does not change over time. The potential is defined piece-wise and therefore again, we have to solve Schrödinger equation piece-wise. We have the three regions:

▷ I: $x < a$
▷ II: $-a < x < a$
▷ III: $x > a$.

We assume the particle in this box has an energy that is lower than V_0, i.e., $E < V_0$. Physically this means that we consider a particle that is *bound* inside this potential[6]. An important new aspect is now that the wave function does not vanish outside of the box. Concretely this means that Ψ is nonzero in the regions I and III. Inside the box we have again the oscillating free solutions that we already discussed in the last section.

$$\Psi_{II}(x) = C \sin\alpha x + D \cos\alpha x \quad -a < x < a,$$

where we have introduced $\alpha \equiv \sqrt{\frac{2mE}{\hbar^2}}$ to unclutter the notation. However, the solutions in the regions I and III now read[7]

$$\begin{aligned}\Psi_I(x) &= A\,\exp^{\beta x} + B\,\exp^{-\beta x} & x < -a \\ \Psi_{III}(x) &= E\,\exp^{\beta x} + F\,\exp^{-\beta x} & x > a,\end{aligned} \quad (7.12)$$

where we introduced a shorthand notation again:

$$\beta \equiv \sqrt{2m(V_0 - E)/\hbar^2}. \quad (7.13)$$

We can check that these functions are indeed solutions in the two regions where the potential is nonzero:[8]

[6] It is always possible to investigate the system for a particle with $E > V_0$ in this case the particle scatters of the box. Such scattering processes are the topic of the next section.

[7] We check this explicitly below.

[8] The check for Ψ_{III} works completely analogous.

$$\frac{d^2}{dx^2}\Psi_I = \frac{2m}{\hbar^2}(V_0 - E)\Psi_I$$

$$\therefore \frac{d^2}{dx^2}\left(A\,\exp^{\beta x} + B\,\exp^{-\beta x}\right) = \frac{2m}{\hbar^2}(V_0 - E)\left(A\,\exp^{\beta x} + B\,\exp^{-\beta x}\right) \quad \text{(using Eq. 7.12)}$$

$$\therefore \beta^2\left(A\,\exp^{\beta x} + B\,\exp^{-\beta x}\right) = \frac{2m}{\hbar^2}(V_0 - E)\left(A\,\exp^{\beta x} + B\,\exp^{-\beta x}\right)$$

$$\therefore \frac{2m(V_0 - E)}{\hbar^2}\left(A\,\exp^{\beta x} + B\,\exp^{-\beta x}\right) = \frac{2m}{\hbar^2}(V_0 - E)\left(A\,\exp^{\beta x} + B\,\exp^{-\beta x}\right) \quad \text{(using Eq. 7.13)} \checkmark$$

These solutions are *non-zero* but still very different from the solution in region II. Take note that there is no imaginary unit in the exponential functions here since $E < V_0$ and thus β is real. An exponential function with the imaginary unit in the argument is an oscillating solution[9], while one without it describes an exponential decay or increase.

[9] Again remember Euler's formula: $e^{ix} = \cos(x) + i\sin(x)$.

A crucial property of wave functions is that they must be normalizable. This means that if we integrate the absolute square

[10] Recall that wave functions describe probability amplitude densities. If we integrate the absolute square of a wave function over some spatial region, we get the probability to find the particle inside this region. If we integrate all over space, the probability to the find the particle must be 100% since the particle has to be somewhere.

of a wave function all over space, the result must be 1 since a probability of more than $1 = 100\%$ doesn't make sense[10]. We can use this immediately to conclude that $B = 0$ and $E = 0$. The reason for this is the following observation: the Region I goes from $-\infty$ to $-a$ and B is the coefficient in front of $\exp^{-\beta x}$. However, $\exp^{-\beta x}$ for $x \to -\infty$ becomes infinite. There is no way to normalize a wave function that becomes infinite somewhere. Therefore B has to be zero. For the same reason, E has to be zero since it is the coefficient in front of $\exp^{\beta x}$ in the region III that goes from a to ∞.

Therefore, the wave function in the region III reads:

$$\Psi_{III}(x) = F \exp^{-\beta x} \tag{7.14}$$

For the moment we are only interested in the general form of this solution. There are two important aspects. Firstly, the solution is nonzero. So while the particle has an energy that is too low to overcome the potential barrier ($E < V_0$) the probability to find it outside the box is non-zero. Classical this is impossible! This phenomenon is usually called **quantum tunneling** or simply tunneling. The particle can tunnel through the potential barrier and get into regions that are classically forbidden.

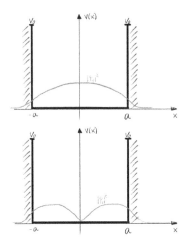

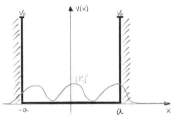

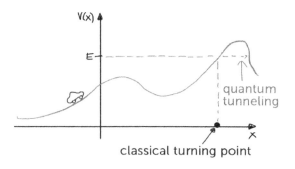

The second important observation is that the function $\exp^{-\beta x}$ describes an exponential decay. Thus while the probability to find the particle outside the box is non-zero, the probability becomes quickly tiny if we move away from the box.

The next thing we can do is to determine the allowed energy values, analogous to what we did in the last section. To do this, we use again the requirement that our wave function and its first derivative must be smooth everywhere[11]

$$\Psi_I(-a) \stackrel{!}{=} \Psi_{II}(-a)$$
$$\partial_x \Psi_I(-a) \stackrel{!}{=} \partial_x \Psi_{II}(-a)$$
$$\Psi_{II}(a) \stackrel{!}{=} \Psi_{III}(a)$$
$$\partial_x \Psi_{II}(a) \stackrel{!}{=} \partial_x \Psi_{III}(a).$$

[11] Reminder: this is necessary since otherwise the momentum and kinetic energy would be infinite.

We are not interested in the most general solution (which is quite complicated) but only in the general features. Therefore, we only consider the case where the wave function inside the box (region II) is symmetric:

$$\Psi_{II}(x) = D \cos \alpha x. \qquad (7.15)$$

In this case, we only have to solve a system of two equations since the symmetry implies that if the function is smooth at a it is also automatically smooth at $-a$. We are therefore left with

$$\Psi_{II}(a) \stackrel{!}{=} \Psi_{III}(a)$$
$$\partial_x \Psi_{II}(a) \stackrel{!}{=} \partial_x \Psi_{III}(a),$$

in which we now put our explicit solutions[12]

$$\Psi_{II}(a) \stackrel{!}{=} \Psi_{III}(a)$$
$$\therefore \quad D \cos \alpha a \stackrel{!}{=} F \exp^{-\beta x} \qquad \text{(using Eq. 7.15 and Eq. 7.14)}$$

and for the second condition which tells us that the first derivatives have to be smooth

$$\partial_x \Psi_{II}(x)\Big|_a \stackrel{!}{=} \partial_x \Psi_{III}(x)\Big|_a$$
$$\therefore \quad \partial_x D \cos \alpha x \Big|_a \stackrel{!}{=} \partial_x F \exp^{-\beta x}\Big|_a$$
$$\therefore \quad -D\alpha \sin \alpha a \stackrel{!}{=} -\beta F \exp^{-\beta a}.$$

[12] We need to be careful what $\partial_x \Psi_{III}(a)$ here means. It means that we first take the derivative and the put in $x = 0$. Otherwise, there is no x dependence and the result would be zero. We indicate this as follows:

$$\partial_x \Psi_{III}(a) = \partial_x \Psi_{III}(x)\Big|_a.$$

This notation makes it clearer that we first take the derivative and only then put in $x = a$.

We divide the second equation by the first one and this yields

$$\alpha \tan \alpha a = \beta \qquad (7.16)$$

This equation determines, analogous to what we discovered in the last section, which energy values are allowed since α and β are both functions of the energy E. Unfortunately it's impossible to solve this equation analytically for E.[13] However, one thing we can do is to solve the equation numerically. Another possibility is to plot the left-hand side and right-hand side separately. The points of intersection are then the allowed solutions.

[13] Mathematicians call this type of equation a **transcendental equation**.

To understand this we define $\xi \equiv a\alpha$ and $\eta \equiv a\beta$ such that Eq. (7.16) now reads

$$\eta = \xi \tan \xi$$

as we can check

$$\eta = \xi \tan \xi$$
$$\therefore \quad a\beta = a\alpha \tan a\alpha$$
$$\therefore \quad \beta = \alpha \tan a\alpha \leftrightarrow \text{Eq. (7.16)} \quad \checkmark$$

We can then plot the allowed values for ξ and η:

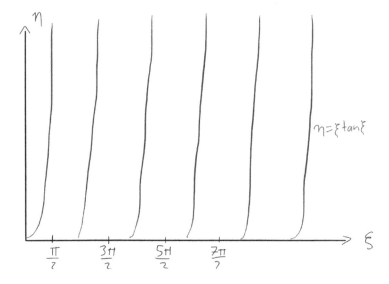

In addition, we have to recall how we defined α and β in the first place

$$\alpha \equiv \sqrt{\frac{2mE}{\hbar^2}}$$

$$\beta \equiv \sqrt{2m(V_0 - E)/\hbar^2}$$

and we can therefore conclude[14]

$$\alpha^2 + \beta^2 \stackrel{!}{=} \frac{2mV_0}{\hbar^2}. \tag{7.17}$$

[14] This condition has to be fulfilled since V_0 is the fixed value that determines how tall our potential walls are.

We can translate this condition into a condition for our newly defined variables η and ξ:[15]

$$\eta^2 + \xi^2 \stackrel{!}{=} \frac{2ma^2V_0}{\hbar^2}$$

$$\therefore \quad (a\beta)^2 + (a\alpha)^2 \stackrel{!}{=} \frac{2ma^2V_0}{\hbar^2}$$

$$\therefore \quad \alpha^2 + \beta^2 \stackrel{!}{=} \frac{2mV_0}{\hbar^2} \leftrightarrow \text{Eq. (7.17)} \quad \checkmark$$

[15] We start here with the final condition and then show why it is true.

This condition for η and ξ defines circles of allowed values. The points where our function $\eta = \xi \tan \xi$ and these circles intersect correspond to the allowed energy values.

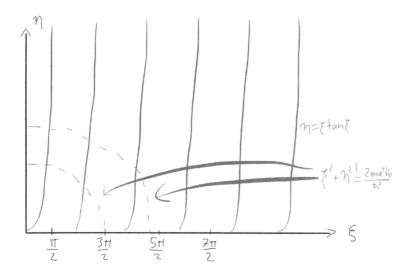

The most important feature is that $\tan x$ is a periodic function and therefore there are multiple solutions. And again, arbitrary values for E are not allowed but only a discrete set. Therefore, the bottom line is: energy is also quantized inside the finite box.

There are dozens of different variations of the box problem. For example, we could investigate how the solution looks like if the box is not symmetric around $x = 0$ and the potential inside the box is nonzero $V = V_0$. However, such changes do not lead to new physical systems. We can always shift our coordinate system and the absolute energy scale such that the system is symmetric and has potential zero inside the box.

Alternatively, we can investigate different shapes of the box or what happens when the box itself moves. A famous example is a box that is infinitesimally thin and infinitely deep (the "Delta potential"). There are no limits and such problems are especially popular in university exams. However, since we do not learn anything qualitatively new in these problems, we do not discuss them here. The most important aspects (tunneling and quantization) are already covered in the two simple examples we discussed in the previous two sections.

The quantum system we discuss in the next section is, in some sense, the masterclass when it comes to quantum systems with bound states. However, the basic lessons are the same that we already discussed in the previous two sections, only the mathematics required is more complicated. For this reason, we will not discuss every detail but only talk about the most important aspects.

7.3 The Hydrogen Atom

Before we start discussing the hydrogen atom there are a few things we should talk about. The thing is that all examples in the previous sections were one-dimensional examples. However, the hydrogen atom is a three-dimensional system and therefore we need to understand what is different if we are dealing with more than one dimension.

First of all, in three dimensions one additional quantity often plays a crucial role: angular momentum.[16] From the discussion in Section 3.4 we know that angular momentum is conserved whenever our system is spherically symmetric. In practice, this means that the potential of the given system is symmetric under rotations. Then it makes sense to switch from Cartesian coordinates (x, y, z) to spherical coordinates (φ, θ, r).

[16] Angular momentum plays no role in 1 dimension, since there are no rotations in 1 dimension.

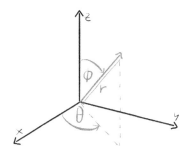

This switch of coordinates is useful since the spherical symmetry means that we only need to consider the variable r for each system individually. The thing is namely that spherical symmetry means that the potential does not depend on φ and θ: $V = V(r)$. Therefore, we first rewrite the stationary Schrödinger equation (Eq. (6.5)) in terms of the new spherical coordinates:

$$-\frac{\hbar^2}{2m}\frac{\partial^2 \psi}{\partial x^2} + V(x)\psi = E\psi \qquad (7.18)$$

$$\therefore \quad -\frac{\hbar^2}{2m}\frac{1}{r}\frac{\partial^2}{\partial r^2}r\psi + \frac{1}{2m}\frac{1}{r^2}\mathbf{L}^2\psi + V(r)\psi = E\psi \qquad (7.19)$$

where[17]

$$\mathbf{L}^2 = -\hbar^2 \left(\frac{1}{\sin\theta}\frac{\partial}{\partial\theta}\sin\theta\frac{\partial}{\partial\theta} + \frac{1}{\sin^2\theta}\frac{\partial^2}{\partial\varphi^2} \right).$$

The trick we then use is the same that we already used in Section 6.1 to derive the stationary Schrödinger equation: separation of the variables.[18] We make the ansatz $\psi(\mathbf{x}) = R(r)Y_{\ell m}(\theta, \varphi)$, where $Y_{\ell m}(\theta, \varphi)$ only depends on the angles and $R(r)$ only the radial distance. Putting this ansatz into the stationary Schrödinger equation yields two equations: one for $Y_{\ell m}(\theta, \varphi)$ and one for $R(r)$. The equations are quite complicated. Solving them requires quite some mathematical machinery and offers little physical insights. So let's only talk about the most important facts.

[17] In fact, this $\mathbf{L}^2$ is exactly the squared angular momentum operator that we already discussed in Section 3.9. The spherical harmonics functions that we talk about below are eigenfunctions of this operator.

[18] In Section 6.1 we used that for systems where the potential does not depend on the time t we can separate the time-dependence and the position-dependence of the wave function: $\Psi(x,t) = \psi(x)T(t)$. Now, we do the same thing again, but this time we separate the angular and radial dependence of the wave function.

The crucial point is that the angular equation (the equation for $Y_{\ell m}(\theta, \varphi)$) is the same for all systems with spherical symmetry since the potential does not depend on the angles. So we only have to solve it once for all these systems. The solutions to the angular part of the Schrödinger equation are known as **spherical harmonics**. These are special functions which are defined by their property that, well, they are solutions of the angular part of the Schrödinger equation.

The radial equation depends on the specifics of the system that we consider and therefore there is no general solution.

The most important example of a system with a spherically symmetric potential is the Hydrogen atom. The relevant potential here is the Coulomb potential $V(r) = -\frac{e^2}{4\pi\varepsilon_0}\frac{1}{r}$. This potential is spherically symmetric since it contains no dependence on the angles θ and φ. Therefore, one part of the solution are the spherical harmonics already mentioned above.

In the solution of the radial Schrödinger equation, a new type of special function appears which is known as **generalized Laguerre polynomials**[19]. An important result is then that the possible energy levels are quantized: $E_n = -\frac{1}{2}m_e \left(\frac{e^2}{4\pi\varepsilon_0\hbar}\right)^2 \frac{1}{n^2}$, completely analogous to what we already discovered for the box examples above.

[19] You can find the details of how these functions and the spherical harmonics are derived in other Quantum Mechanics textbooks. I'm sure if you see the calculation you will be able to follow the steps. However, it is also clear that it would take a lot of time to come up with these solutions on your own. Historically, these weren't discovered in an hour either. So don't worry, students are usually not expected to come up with solutions for such complicated systems. Later if you do actual research, it is possible that you have to find solutions of the Schrödinger equation for comparably complicated systems. However, there is no general method to come up with solutions. Usually the best idea is to ask a trustworthy colleague from the math department. This is why we don't discuss all the details here.

8

Scattering off a Box

In the previous sections we only discussed **bound states**. Such states describe particles which are trapped inside some potential.

The second type of state that exists in Quantum Mechanics are **scattering states**[1]. These states describe particles that move, for example, towards some potential barrier.

[1] As discussed in Section 6.3, bound states are characterized by $E < V(\infty)$ and $E < V(-\infty)$, while scattering states are characterized by $E > V(\infty)$ or $E > V(-\infty)$.

Depending on their energy and the height of the potential barrier the particles either get reflected, or pass through or above the barrier. As always in Quantum Mechanics each such process is possible with a certain probability.

So the situation we are now interested in looks like this[2]

[2] We consider here a particle that moves from the left towards the potential barrier. Of course, it is also possible to consider a particle that comes from the right.

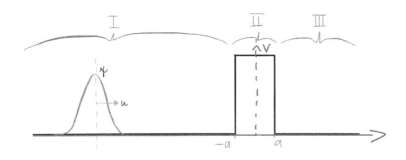

Then, after the interaction with the potential barrier, the situation looks like this

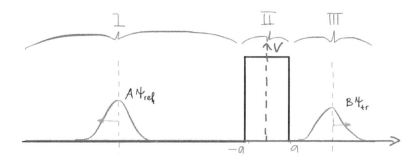

As in the previous section we have to solve the Schrödinger equation piece-wise because the potential is defined piece-wise. In the Regions I and II we have no potential and therefore the usual free particle solutions:

$$\begin{aligned}\Psi_I(x) &= \exp^{ikx} + A\exp^{-ikx} & x < -a \\ \Psi_{III}(x) &= B\exp^{ikx} & x > a,\end{aligned} \quad (8.1)$$

with $k \equiv \sqrt{2mE/\hbar^2}$. Take note that there is no coefficient in front of $\exp^{ikx}$ in Region I and no function of the form $\exp^{-ikx}$ in Region III since we consider a particle that moves from the left towards the potential barrier[3]. In other words, we do not consider the most general system, but a very specific physical situation.

[3] We will talk about this in more detail in a moment.

The probability that our particle passes through the barrier or gets reflected are directly related to the coefficients A and B.

$$P(\text{reflection}) = |A|^2$$
$$P(\text{transmission}) = |B|^2.$$

Therefore, our only task is to calculate these coefficients for different potential barriers and different energies of the incoming particle. The energy of the particle (relative to the height of the potential barrier) is important since if it is high enough the particle can pass above the barrier. If the energy is lower than the potential barrier, the particle has to tunnel through the barrier. So the solution in the Region II depends on whether the energy is higher or lower than the potential barrier. If it is *higher* we also get here an *oscillating* solution

$$\Psi_{II}(x) = C\exp^{i\kappa x} + D\exp^{-i\kappa x} \qquad -a < x < a, \qquad (8.2)$$

but with a modified frequency $\kappa \equiv \sqrt{2m(E-V)/\hbar^2}$.

If the energy is *lower* than the potential barrier $E < V$, we have something negative in the square root here: $\sqrt{2m(E-V)/\hbar^2}$. The square root of something negative is imaginary. In such a situation we factor out a -1 from the square root, which then becomes[4]

$$\kappa = \sqrt{2m(E-V)/\hbar^2} = \sqrt{(-1)2m(V-E)/\hbar^2}$$
$$= \sqrt{(-1)}\sqrt{2m(V-E)/\hbar^2} = i\sqrt{2m(V-E)/\hbar^2} \equiv \tilde{\kappa}.$$

[4] We will discuss all this more explicitly in the following sections.

The solution in Region II then reads

$$\Psi_{II}(x) = C\exp^{-\tilde{\kappa}x} + D\exp^{\tilde{\kappa}x} \qquad -a < x < a. \qquad (8.3)$$

So we get an exponential decay of the wave function, which means physically that the particle tunnels through the barrier.

Before we discuss the most important examples, a few comments that hopefully answer the most pressing questions you probably have at this point:

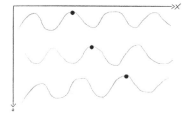

▷ One thing I was confused about as a student is why e^{ix} describes a wave that moves from left to right. This interpretation only makes sense if we remember that we consider here solutions $\psi(x)$ of the *stationary* Schrödinger equation. These solutions do not move at all! The full solutions are $\Psi(x,t) = T(t)\psi(x) = \psi(x)\exp\left(-\frac{iEt}{\hbar}\right)$ (Eq. (6.6)). So for a solution of the form $\psi(x) = e^{ix}$ (ignoring $\hbar$ for a moment), we have $\Psi(x,t) = e^{ix-iEt}$. Now, let's focus on one specific point in our incoming wave. We choose the point which is at $t=0$ at the location $-\pi$. The value of the wave function at this point is $\Psi(x=-\pi, t=0) = \exp(-i\pi) = -1$. Now, we want to understand how the wave function moves. As time passes on t becomes larger. This, in turn, means that the point that we picked is now at a larger x since there is a minus sign between the x and t part in the exponent. As x gets larger, we move on the x-axis from the left to the right. This shows that the wave really moves from the left to the right.

▷ Another question that usually comes up at this point is why there is no coefficient in front of the incoming wave. This is a result of our physical choice that the particle moves towards the potential from the left. In words this means that 100% of the particles come from here. However, take note that the total probability is still not larger than 100%. The coefficients are only probability *amplitudes*. Only the absolute square of them is related to probability. Especially, take note that we calculate the probability to find the particle in region I by taking the absolute square of the corresponding wave function. In general, the result will not be 1. Physically this means that there is some probability for the particle to move beyond the potential barrier.

▷ In the images above, we have wave packets that move towards and away from the potential barrier. However, as already explained at the beginning of this part of the book, calculations with wave packets are, in general, too difficult. Therefore in the following, we talk about the simpler plane waves instead (Eq. (8.1)). This is a valid approach since wave packets consists of plane waves[5].

[5] Formulated differently: a wave packet is a superposition of many plane waves.

8.1 The Step Potential

The first concrete potential we now take a look at is a box that spreads out infinitely to the right:

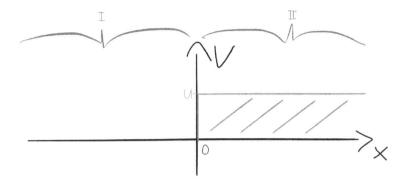

or mathematically

$$V(x) = \begin{cases} 0 & x \leq 0 \\ U & x > 0 \end{cases},$$

where $U > 0$ is a constant. So it is not really a box potential but rather a **step potential**. As discussed in the introduction we consider a particle that moves towards the potential from the left.

Once more our task is to solve the stationary Schrödinger equation (Eq. (6.5))[6]

$$\frac{-\hbar^2}{2m}\psi'' + V(x)\psi = E\psi.$$

Since the potential is defined piece-wise we have to solve it piece-wise. A crucial idea to connect the solutions in the different regions is again that our wave function and its first derivative are smooth everywhere[7].

As already mentioned above there are two different cases we can discuss. Either the incoming particle has a large enough energy to overcome the barrier $E > U$, or it is too small $E < U$. In the second case, classically the particle would definitely get reflected, but in Quantum Mechanics the particle can also tunnel through the barrier. So there is a nonzero probability to find it in Region II.

We discuss both cases in the following two sections.

[6] We use the shorthand notation $\psi' \equiv \partial_x \psi$ and $\psi'' \equiv \partial_x^2 \psi$.

[7] Recall that the physical reason for this requirements is that otherwise the momentum or kinetic energy would be infinite.

8.1.1 $E < U$

As already discussed in the introduction, we define two new constants to simplify the notation

$$k \equiv \sqrt{2mE/\hbar^2}$$
$$\tilde{\kappa} \equiv \sqrt{2m(U-E)/\hbar^2}$$

The Schrödinger equation in the two regions then reads

$$\begin{cases} \psi'' + k^2\psi = 0 & x < 0 \\ \psi'' - \tilde{\kappa}^2\psi = 0 & x > 0. \end{cases}$$

The corresponding solutions are the usual free particle solution in Region I: $\psi = 1e^{ikx} + Ae^{-ikx}$, and the tunnel solution $\psi = Be^{-\tilde{\kappa}x}$ in Region II. Take note that in principle we also have a term of the form $e^{\tilde{\kappa}x}$ in Region II. However, this term is not allowed since otherwise the wave function becomes infinitely large for $x \to \infty$.

Next, we demand that ψ and ψ' are continuous at $x = 0$ and this yields

$$\begin{cases} 1 + A = B \\ ik1 - ikA = -\tilde{\kappa}B. \end{cases}$$

Therefore, we find[8]

$$A = \frac{k - i\tilde{\kappa}}{k + i\tilde{\kappa}}, \quad B = \frac{2k}{k + i\tilde{\kappa}}.$$

Now, if someone gives you explicit values for the potential barrier height U and the energy of the particle E you could immediately calculate the probabilities to find the particle behind or before the barrier explicitly.

An important observation is that for $U \to \infty$ we have $|B| \to 0$ since $\tilde{\kappa} = \sqrt{2m(U-E)/\hbar^2} \to \infty$. Physically this means that for an infinitely high potential barrier no tunneling is possible.

In the next section we discuss the same system again, but this time the particle's energy is higher than the potential barrier.

---

[8] The equations above are a system of two equations for the two unknowns A and B. We do not discuss here how we solve such a system of equations since it has nothing to do with Quantum Mechanics. If you're unsure how to solve such a system you can use, for example, http://wolframalpha.com.

8.1.2 $E > U$

For this situation we define

$$k \equiv \sqrt{2mE/\hbar^2}$$
$$\kappa \equiv \sqrt{2m(E-U)/\hbar^2}$$

and the Schrödinger equation then reads

$$\begin{cases} \psi'' + k^2\psi = 0 & x < 0 \\ \psi'' + \kappa^2\psi = 0 & x > 0 \end{cases}$$

Again, we have in Region I ($x < 0$) the solution $\psi = e^{ikx} + R^{-ikx}$. But this time we have in Region II ($x > 0$) the oscillating solution $\psi = Te^{ikx}$.[9] Take note that, in principle, we could also have an e^{-ikx} term in Region II. However, this would correspond to a particle that moves towards the potential barrier from the right. The physical situation that we investigate here has only a particle that moves towards the barrier from the left side.

[9] Take note that T is here a constant and not the time-dependent part of the full Schrödinger equation.

Next, we again use the fact that ψ and ψ' must be smooth at $x = 0$. This yields the equations

$$\begin{cases} I + R = T \\ ikI - ikR = ikT. \end{cases}$$

and solving them yields

$$R = \frac{k-\kappa}{k+\kappa}I, \quad T = \frac{2k}{k+\kappa}I.$$

In the previous section, we had a particle with too little energy to overcome the barrier classically. However, in Quantum Mechanics there is some probability that the particle tunnels through the barrier nevertheless. Now we have a particle with sufficient energy to overcome the potential barrier. Classically the particle would therefore simply move beyond the barrier. However, in Quantum Mechanics there is a nonzero probability that the particle gets reflected even though it has enough energy.

8.2 The Box Potential

The next system we have a look at is very similar to the previous one. The only difference is that the potential barrier does not extend infinitely to the right:

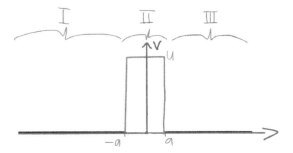

So mathematically, we have

$$V(x) = \begin{cases} 0 & x \leq 0 \\ U & 0 < x < a \\ 0 & x \geq a \end{cases}$$

While it is also possible to discuss the two cases $E < U$ and $E > U$, here we restrict ourselves to the former case here since for the latter case we wouldn't learn anything new here.

Again, we define

$$k \equiv \sqrt{2mE/\hbar^2}$$
$$\tilde{\kappa} \equiv \sqrt{2m(U-E)/\hbar^2}.$$

The Schrödinger equation in the three regions then reads

$$\begin{cases} \psi'' + k^2\psi = 0 & x < 0 \\ \psi'' - \tilde{\kappa}^2\psi = 0 & 0 < x < a \\ \psi'' - \tilde{\kappa}^2\psi = 0 & x > a. \end{cases}$$

The solutions are of the following form which is somewhat familiar by now:

$$\psi = e^{ikx} + Re^{-ikx} \qquad x < 0$$
$$\psi = Ae^{\kappa x} + Be^{-\kappa x} \qquad 0 < x < a$$
$$\psi = Te^{ikx} \qquad x > a.$$

Using the smoothness of ψ and ψ' at $x = 0$ and a gives us the equations

$$1 + R = A + B$$
$$ik(1 - R) = \kappa(A - B)$$
$$Ae^{\kappa a} + Be^{-\kappa a} = Te^{ika}$$
$$\kappa(Ae^{\kappa a} - Be^{-\kappa a}) = ikTe^{ika}.$$

We solve these and obtain[10]

$$1 + \frac{\kappa - ik}{\kappa + ik}R = Te^{ika}e^{-\kappa a}$$
$$1 + \frac{\kappa + ik}{\kappa - ik}R = Te^{ika}e^{\kappa a}.$$

After a long and tedious calculation we find

$$T = e^{-ika}\left(\cosh \kappa a - i\frac{k^2 - \kappa^2}{2k\kappa}\sinh \kappa a\right)^{-1}.$$

The most important fact is that T is in general nonzero. This once more demonstrates the phenomenon of quantum tunneling explicitly.

[10] Again, we do not discuss here how we solve such equations since this is purely a math problem and does not help us to understand Quantum Mechanics any better.

Again, it's of course possible to study arbitrarily complicated variations of the problems we discussed in the previous two sections.

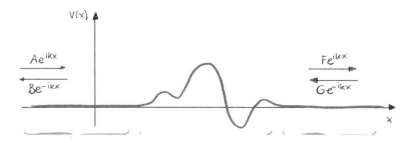

However, the basic method is always the same. And most importantly, Quantum Mechanics is not really to right theory to study interesting scattering processes. In interesting scattering processes multiple particles collide and new particles are created through the scattering. It is extremely difficult to study such processes in Quantum Mechanics and the right theory to deal with this kind of problems is Quantum Field Theory.

Now, we move on to another incredibly famous and important quantum system: the quantum harmonic oscillator.

9

Harmonic Quantum Mechanics

Sidney Coleman once remarked that *"The career of a young theoretical physicist consists of treating the harmonic oscillator in ever-increasing levels of abstraction."*

So it certainly makes sense to study the quantum harmonic oscillator in some detail. But first, what is a harmonic oscillator?

The answer is: All systems with a potential of the form $V = cx^2$, where c is some constant. Plotted the potential of the harmonic oscillator looks like this

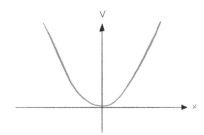

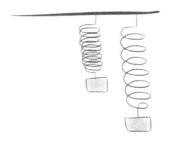

An example of such a system is an object attached to a spring.

And why are harmonic oscillators so important?

The thing is that in a first approximation *lots* of potentials are extremely similar to the harmonic potential. Concretely this means that the first term of the Taylor expansion[1] of many potentials is exactly the harmonic potential:

[1] For the basic idea behind the Taylor expansion see Appendix A.

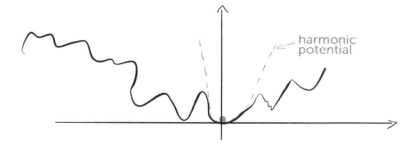

For example, the Taylor expansion of a much more complicated function like $\cos x$ is

$$\cos x = 1 - \frac{x^2}{2} + \ldots.$$

So for small x the potential can be approximated by $1 - x^2$. A concrete physical example is a pendulum which is described by the potential $V = 1 - \cos x$:

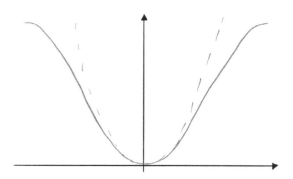

Thus, by studying the harmonic potential closely we can learn a lot about lots of other systems, at least as long as we are dealing with small excitation (low energies).

The potential of the harmonic oscillator is usually written as

$$V(x) = \frac{1}{2}kx^2, \qquad (9.1)$$

where k is the **spring constant** which characterizes the strength of the spring. Alternatively, we write the potential often as

$$V(x) = \frac{1}{2}m\omega^2 x^2, \qquad (9.2)$$

where $\omega = \sqrt{k/m}$ denotes the classical oscillation frequency and m the mass at the end of the spring.

The stationary Schrödinger equation (Eq. (6.5)) therefore reads

$$-\frac{\hbar^2}{2m}\frac{\partial^2 \psi}{\partial x^2} + V(x)\psi = E\psi$$

$$\therefore \quad -\frac{\hbar^2}{2m}\partial_x^2 \psi + \frac{1}{2}m\omega^2 x^2 \psi = E\psi \quad \text{(using Eq. 9.2)} \qquad (9.3)$$

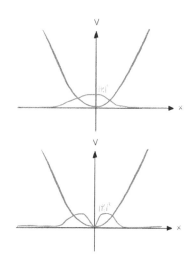

Figure 9.1: Wave functions for the harmonic potential. Take note how similar these are to the solutions of the finite box system.

Now, Eq. (9.3) is a quite complicated differential equation. It's possible to solve it using clever mathematical tricks. However, our goal here is to understand Quantum Mechanics and not to solve complicated equations. Only for completeness let me mention that the solutions to Eq. (9.3) look like this

$$\psi_n(x) = (\frac{1}{2})^{n/2} H_n\left(\sqrt{\frac{m\omega}{\hbar}}x\right) e^{-\frac{m\omega}{2\hbar}x^2},$$

where H_n denotes the so-called **Hermite polynomials**

$$H_n(u) = (-1)^n e^{u^2/2} \frac{d}{du} e^{-u^2/2}.$$

The first few Hermite polynomials are:

$H_0(u) = 1, \quad H_1(u) = 2u, \quad H_2(u) = 4u^2 - 2, \quad H_3(u) = 8u^3 - 12u.$

The functions $\Psi_n(x)$ are the energy eigenstates since they are solutions to the stationary Schrödinger equation. The corresponding energy eigenvalues are

$$E_n = \hbar\omega\left(n + \frac{1}{2}\right).$$

And once more: these functions describe the probability amplitude densities to find the particle in certain spatial regions. So in principle, this is all we want to know. We know the energies, and we can calculate where we will most probably find the particle if the particle has energy E_n, e.g., energy E_1.

There are two important lessons here:

▷ Quantum Mechanics gets complicated quickly. Anything beyond simple boxes is almost impossible to solve analytically.

▷ We need a smart idea to get any deeper understanding of the quantum harmonic oscillator.

Luckily, there is an incredibly smart method that makes the whole problem much more transparent. This method is the topic of the next section.

9.1 The Magical Method

As promised, here is the smart idea: We define the following two operators and then use them instead of $\hat{x}$ and $\hat{p}$[2]

[2] Take note that $a^\dagger$ is the Hermitian adjoint of a, where $\dagger \equiv *T$, i.e., conjugation plus transposition.

$$a \equiv \sqrt{\frac{m\omega}{2\hbar}}x + i\frac{1}{\sqrt{2m\omega\hbar}}p \qquad (9.4)$$

$$a^\dagger \equiv \sqrt{\frac{m\omega}{2\hbar}}x - i\frac{1}{\sqrt{2m\omega\hbar}}p. \qquad (9.5)$$

We will understand the physical meaning of these operators in a moment.

We can also invert these equations:

$$\text{Eq. 8.4 + Eq. 8.5} \Rightarrow a + a^\dagger = 2\sqrt{\frac{m\omega}{2\hbar}}x$$

$$\therefore \quad x = \sqrt{\frac{\hbar}{2m\omega}}(a + a^\dagger) \qquad (9.6)$$

and

$$\text{Eq. 8.4 + Eq. 8.5} \Rightarrow a - a^\dagger = 2i\frac{1}{\sqrt{2m\omega\hbar}}p$$

$$\therefore \quad p = -i\sqrt{\frac{\hbar m\omega}{2}}(a - a^\dagger). \qquad (9.7)$$

In addition, using the canonical commutation relation $[x, p] = i\hbar$ we can calculate the commutator of a and $a^\dagger$:[3]

[3] It will become clear in a moment why this is useful.

$$[a, a^\dagger] = aa^\dagger - a^\dagger a$$
$$= \left(\sqrt{\frac{m\omega}{2\hbar}}x + i\frac{1}{\sqrt{2m\omega\hbar}}p\right)\left(\sqrt{\frac{m\omega}{2\hbar}}x - i\frac{1}{\sqrt{2m\omega\hbar}}p\right) \quad \text{(using Eq. 9.4 and Eq. 9.5)}$$
$$\quad - \left(\sqrt{\frac{m\omega}{2\hbar}}x - i\frac{1}{\sqrt{2m\omega\hbar}}p\right)\left(\sqrt{\frac{m\omega}{2}}x + i\frac{1}{\sqrt{2m\omega\hbar}}p\right)$$
$$= \frac{m\omega}{2\hbar}x^2 - \frac{i}{2\hbar}xp + \frac{i}{2\hbar}px + \frac{1}{2m\omega\hbar}p^2 - \frac{m\omega}{2\hbar}x^2 - \frac{i}{2\hbar}xp + \frac{i}{2\hbar}px - \frac{1}{2m\omega\hbar}p^2$$
$$= \frac{i}{\hbar}(px - xp) = \frac{i}{\hbar}[p, x] = -\frac{i}{\hbar}[x, p] = -\frac{i}{\hbar}i\hbar$$
$$= 1 \qquad (9.8)$$

We can now use these equations to rewrite the Schrödinger equation for the harmonic oscillator (Eq. (9.3)) in terms of a and $a^\dagger$

$$E\Psi = \frac{p^2}{2m}\Psi + \frac{m\omega^2}{2}x^2\Psi$$
$$= \frac{1}{2m}\left(i\sqrt{\frac{\hbar m\omega}{2}}(a^\dagger - a)\right)^2\Psi + \frac{m\omega^2}{2}\left(\sqrt{\frac{\hbar}{2m\omega}}(a + a^\dagger)\right)^2\Psi \quad \text{(using Eq. 9.6 and Eq. 9.7)}$$
$$= \frac{\hbar\omega}{4}\left(-a^\dagger a^\dagger + a^\dagger a + aa^\dagger - aa\right)\Psi + \frac{\hbar\omega}{4}\left(aa + a^\dagger a + aa^\dagger + a^\dagger a^\dagger\right)\Psi$$
$$= \frac{\hbar\omega}{2}\left(a^\dagger a + aa^\dagger\right)$$

We can then use the commutator relation in Eq. (9.8) to rewrite this as follows

$$E\Psi = \frac{\hbar\omega}{2}\left(a^\dagger a + a a^\dagger\right)$$
$$= \frac{\hbar\omega}{2}(a^\dagger a + a a^\dagger - a^\dagger a + a^\dagger a) \quad (-a^\dagger a + a^\dagger a = 0)$$
$$= \frac{\hbar\omega}{2}(2a^\dagger a + [a, a^\dagger])$$
$$= \frac{\hbar\omega}{2}\left(2a^\dagger a + 1\right) \quad ([a, a^\dagger] = 1, \text{Eq. 9.8})$$
$$= \hbar\omega\left(a^\dagger a + \frac{1}{2}\right)$$

What we have here on the right-hand side is the quantum energy operator. As already mentioned we usually call this operator the Hamiltonian and denote it by H:

$$H \equiv \hbar\omega\left(a^\dagger a + \frac{1}{2}\right). \qquad (9.9)$$

Acting with this operator on a state that describes our system tells us the energy of the system[4]:

$$H\,|E_1\rangle = E_1\,|E_1\rangle. \qquad (9.10)$$

[4] At least, when the system is in an energy eigenstate.

Now, we want to understand these new operators a and $a^\dagger$. The most important thing for us is what a and $a^\dagger$ do when they act on our system: $a\,|E_1\rangle=?$ To get a feeling for this, let's calculate the energy of such a new state[5]. Before we can do this, we need one more thing: the commutator $[H, a]$ as we will see in a second. Using Eq. (9.8) and Eq. (9.9), we find

[5] Acting with a on our state $|E_1\rangle$ yields a new state. What we do now is to check if they operators change the energy. All this will make a lot more sense in a moment.

$$[H, a] = Ha - aH$$
$$= \left(\hbar\omega\left(a^\dagger a + \frac{1}{2}\right)\right)a - a\left(\hbar\omega\left(a^\dagger a + \frac{1}{2}\right)\right) \quad \text{(using Eq. 9.9)}$$
$$= \hbar\omega\left(a^\dagger a a + \frac{a}{2} - a a^\dagger a - \frac{a}{2}\right)$$
$$= \hbar\omega\left(a^\dagger a - a a^\dagger\right)a$$
$$= \hbar\omega[a^\dagger, a]a$$
$$= -\hbar\omega[a, a^\dagger]a \quad ([a, a^\dagger] = -[a^\dagger, a])$$
$$= -\hbar\omega a \quad ([a, a^\dagger] = 1, \text{Eq. (9.8)}) \qquad (9.11)$$

Completely analogous, we can calculate

$$[H, a^\dagger] = \hbar\omega a^\dagger. \qquad (9.12)$$

HARMONIC QUANTUM MECHANICS 143

With this information at hand, we are finally ready to calculate the energy of our new state $a\,|E_1\rangle$:

$$\hat{H}(a\,|E_1\rangle) = (\hat{H}a - a\hat{H} + a\hat{H})\,|E_1\rangle \qquad (-a\hat{H} + a\hat{H} = 0)$$
$$= a\hat{H}\,|E_1\rangle + [\hat{H}, a]\,|E_1\rangle \qquad (\hat{H}a - a\hat{H} \equiv [\hat{H},a])$$
$$= aE_1\,|E_1\rangle + [\hat{H}, a]) \qquad \text{(using Eq. 9.10)}$$
$$= \left(aE_1 - \hbar\omega a\right)|E_1\rangle \qquad \text{(using Eq. 9.11)}$$
$$= \left(E_1 - \hbar\omega\right)\left(a\,|E_1\rangle\right). \qquad (9.13)$$

Analogously for $a^\dagger$ we find

$$\hat{H}a^\dagger\,|E_1\rangle = (E_1 + \hbar\omega)a^\dagger\,|E_1\rangle. \qquad (9.14)$$

What do we learn here?

By looking at Eq. (9.13) we see that $a\,|E_1\rangle$ can be interpreted as a new state with energy $E - \hbar\omega_k$!

Let's make this more concrete. We define

$$|E_0\rangle \equiv a\,|E_1\rangle \qquad (9.15)$$

with

$$\hat{H}|E_0\rangle = \hat{H}(a\,|E_1\rangle) \qquad \text{(using Eq. 9.15)}$$
$$= (E_1 - \hbar\omega)\,|E_0\rangle \qquad \text{(using Eq. 9.13)}$$

Analogously, by looking at Eq. (9.14) we see that $a^\dagger\,|E_1\rangle$ can be interpreted as a new state with energy $E + \hbar\omega_k$.

Concretely, we have

$$|E_2\rangle \equiv a^\dagger\,|E_1\rangle$$

with

$$\hat{H}\,|E_2\rangle = (E_1 + \hbar\omega)\,|E_2\rangle \qquad \text{(using Eq. 9.14)}.$$

This is why a and $a^\dagger$ are known as **ladder operators**. They allow us to move between the energy eigenstates. Using $a^\dagger$ we can jump to the next higher eigenstate. Using a we can jump to the next lower eigenstate.

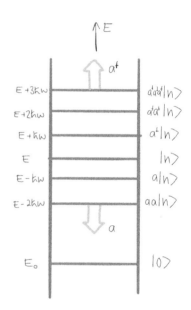

The non-trivial part of the Hamiltonian operator (Eq. (9.9)) is

$$N \equiv a^\dagger a. \tag{9.16}$$

With this definition, we can rewrite the Hamiltonian in the form:

$$H = \hbar\omega\left(N + \frac{1}{2}\right). \tag{9.17}$$

So the operator N tells us in which energy eigenstate we currently are:

$$N|n\rangle = n|n\rangle, \tag{9.18}$$

where we use n to label the n-th energy eigenstate with energy E_n. It is conventional to use the notation $|1\rangle, |2\rangle, \ldots, |n\rangle$ for these states. The operators a and $a^\dagger$ let us move between them in discrete jumps.

One important thing we need to take care of is the following:

States in Quantum Mechanics always need to be normalized. Otherwise, our probabilistic interpretation makes no sense since probabilities of more than 100% do not make sense. However, when we act with an operator on some state, we can not expect that the new state we generate this way is automatically normalized, too. Instead, in general, we get something of the form

$$a^\dagger|n\rangle = C|n+1\rangle, \tag{9.19}$$

where $|n\rangle$ and $|n+1\rangle$ are normalized states. In other words, when we act with $a^\dagger$ or a on a ket $|n\rangle$, we do not simply get $|n\pm1\rangle$ but possibly an additional constant factor C. So to determine completely what a and $a^\dagger$ do, we need to determine these constant factors. We can do this by using that $|n\rangle$ and $|n\pm1\rangle$ are normalized. We first calculate[6]

$$a^\dagger|n\rangle)^\dagger = (C|n+1\rangle)^\dagger$$
$$\to \langle n|a = \langle n+1|C^\dagger. \tag{9.20}$$

[6] Recall that $\langle\Psi| \equiv |\Psi\rangle^\dagger$.

and can then calculate

$$\langle n|\,aa^\dagger\,|n\rangle = \langle n+1|C^\dagger C|n+1\rangle \quad \text{(using Eq. 9.20 and Eq. 9.19)}$$
$$= C^\dagger C \underbrace{\langle n+1|n+1\rangle}_{=1} \quad \text{(C is a number, not an operator)}$$
(9.21)

Alternatively, we can evaluate this expression using the commutation relation in Eq. (9.8)

$$\langle n|\,aa^\dagger\,|n\rangle = \langle n|\left(aa^\dagger - a^\dagger a + a^\dagger a\right)|n\rangle \quad (-a^\dagger a + a^\dagger a = 0)$$
$$= \langle n|\left([a,a^\dagger] + a^\dagger a\right)|n\rangle$$
$$= \langle n|\left(1 + a^\dagger a\right)|n\rangle \quad \text{(using Eq. 9.8)}$$
$$= \langle n|\left(1 + N\right)|n\rangle \quad \text{(using Eq. 9.16)}$$
$$= \langle n|\left(N|n\rangle + 1|n\rangle\right)$$
$$= \langle n|\left(n|n\rangle + 1|n\rangle\right) \quad \text{(using Eq. 9.18)}$$
$$= \langle n|\,(n+1)\,|n\rangle$$
$$= (n+1)\langle n|n\rangle \quad (n+1 \text{ is a number, not an operator})$$
$$= n+1 \quad (\langle n|n\rangle = 1). \quad (9.22)$$

Putting Eq. 9.21 and Eq. 9.22 together yields

$$\langle n|\,aa^\dagger\,|n\rangle = \langle n|\,aa^\dagger\,|n\rangle$$
$$\therefore\ C^\dagger C = n+1$$
$$\therefore\ C = \sqrt{n+1}. \quad (9.23)$$

We can therefore conclude

$$a^\dagger |n\rangle = C|n+1\rangle \quad \text{(this is Eq. 9.19)}$$
$$= \sqrt{n+1}|n+1\rangle \quad \text{(using Eq. 9.23)} \quad (9.24)$$

Following the same steps we can derive

$$a|n\rangle = \sqrt{n}|n-1\rangle. \quad (9.25)$$

This result is important because it tells us that the ladder ends. For $n = 0$ we get

$$a|0\rangle = \sqrt{0}|0-1\rangle = 0. \quad (9.26)$$

This means that if we act with the **lowering operator** a on the state with label 0, we don't get a new state but simply 0. So this is where the ladder ends. In physical terms this means that there is a state with minimum energy. Acting with $H = \hbar\omega\left(N + \frac{1}{2}\right)$ on this state yields

$$H\left|0\right\rangle = \hbar\omega\left(N + \frac{1}{2}\right)\left|0\right\rangle$$

$$= \hbar\omega\left(a^\dagger a + \frac{1}{2}\right)\left|0\right\rangle \quad (N \equiv a^\dagger a, \text{Eq. 9.16})$$

$$= \hbar\omega\left(0 + \frac{1}{2}\right)\left|0\right\rangle \quad (a\left|0\right\rangle = 0, \text{Eq. 9.26})$$

$$= \frac{\hbar\omega}{2}\left|0\right\rangle.$$

This means the ground state energy of the quantum harmonic oscillator is $E_0 = \hbar\omega/2$.

Let's summarize what we have learned here:

▷ There is a state with the lowest possible energy $E_0 = \frac{\hbar\omega}{2}$. This is the **ground state energy** of the Harmonic oscillator. Interestingly it is non-zero, so there is always *some* fluctuation.[7]

▷ All other energy eigenstates can be generated by acting with the **raising operator** $a^\dagger$ on this state with the lowest energy multiple times. Each time we use $a^\dagger$, we generate a new state with an energy that is $\hbar\omega$ higher than the previous one.

▷ The energy spectrum is therefore again *discrete*. The distance between the energy states is $\hbar\omega$.

---

[7] This leads to a curious result in Quantum Field Theory. As already mentioned above, in some sense, a quantum field is a set of infinitely many harmonic oscillators. Now, we just learned that the ground state energy of a single harmonic oscillator is nonzero, but $E_0 = \frac{\hbar\omega}{2}$. The ground state energy of system consisting of two harmonic oscillators is therefore $\hbar\omega$, and the ground state energy of a system consisting of infinitely many harmonic oscillators is infinity. In other words, the ground state energy of a quantum field is infinity. However, we can only measure energy differences anyway and since every field has an infinite ground state energy, we usually simply ignore this infinite result.

10

Quantum Systems with Spin

In all examples we discussed so far, we assumed that the particles are **structureless**. This means that we assumed that it is sufficient to describe them using the variables: position, energy, and momentum.

However, as we already discussed in Section 3.10, certain particles have some kind of internal structure known as **spin**.[1] This property is not always relevant and only makes a difference for specific systems. Formulated differently, often we can ignore that our particles have spin, at least in a first approximation.

[1] Actually all known elementary particles have a nonzero spin, except for the Higgs boson.

But there are, of course, also systems where a nonzero spin makes all the difference.

To understand this a bit better imagine the following situation:

You throw a ping-pong ball to a friend who stands in front of you. The ball is perfectly round and perfectly white. You then ask your friend: "was the ball spinning as it was traveling towards you?" Since the ball is perfectly round and perfectly white, there is no way he can answer this question. In addition, the trajectory of the ball does not depend on whether it spins or not - at least as long as we neglect air resistance. We can calculate the trajectory using Newton's second law $F = ma$ plus Newton's law for gravity $F = -mgh$ and there is nothing that changes if the ball spins. However, the internal spinning becomes incredibly important if we change the setup a bit. For example, if the ball collides with another ball. Another example is when the ball is larger such that interactions with the air surrounding it makes a significant difference[2].

In the examples above we neglected that elementary particles have spin. We already mentioned a system where spin is important in Section 3.10: the Stern-Gerlach experiment. The analogue to our ping-pong ball is here an electron (or analogously a silver atom). The analogue to the air surrounding our spinning ping-pong ball is a magnetic field. Due to the presence of the magnetic field, it makes a huge difference how exactly our electron spins. When an electron spins, it behaves like a tiny magnet since it carries electric charge. Completely analogous to how a magnet gets deflected in a magnetic field, now our spinning electron gets deflected[3].

As a reminder of what we already discussed in Section 3.10: the spin operators for an electron look like this[4]

$$\hat{S}_i = \frac{\hbar}{2}\sigma_i, \qquad (10.1)$$

where σ_i are the Pauli matrices[5]. One thing we can see immediately here is that no matter how we measure the spin of an electron the result is always $\hbar/2$ or $-\hbar/2$, which we usually call "spin up" and "spin down". This follows since the eigenvalues of each of the three spin operator matrices are exactly $\hbar/2$ or $-\hbar/2$.[6] In physical terms, this means that no matter whether we measure the spin along the z-axis, the x-axis, or along the y-axis, the result is always either $\hbar/2$ or $-\hbar/2$. There are only two possible outcomes since the spin operators are here given by (2×2) matrices.

We now consider a concrete example of how a spin-measurement works in the quantum formalism. We will not only discover one critical feature of Quantum Mechanics, but also get a deeper un-

---

[2] Just think about how big the difference is between the trajectories of a spinning football and one that does not spin.

[3] The electron has an *intrinsic* magnetic dipole moment $\vec{M}$ due to its spin, which is given by $\vec{M} = -eg_S\vec{S}/2m$, where $g_S = 2(1 + \alpha/2\pi + \cdots)$ denotes the gyromagnetic ratio. So if there is a an external magnetic field B the electron also has a potential energy $U = -\vec{M} \cdot \vec{B}$ which we need to include in the Schrödinger equation.

[4] We have multiple spin operators. Each measures the spinning around a different axis.

[5]
$$\sigma_1 = \begin{pmatrix} 0 & 1 \\ 1 & 0 \end{pmatrix},$$
$$\sigma_2 = \begin{pmatrix} 0 & -i \\ i & 0 \end{pmatrix},$$
$$\sigma_3 = \begin{pmatrix} 1 & 0 \\ 0 & -1 \end{pmatrix}.$$

[6] We discussed eigenvalues in Section 3.8.

derstanding of how the quantum formalism works in practice.

10.1 Spin Measurements

The eigenstates of the $\hat{S}_z$ operator

$$\hat{S}_z = \frac{\hbar}{2}\sigma_3 = \frac{\hbar}{2}\begin{pmatrix} 1 & 0 \\ 0 & -1 \end{pmatrix} \quad (10.2)$$

are

$$|\hbar/2\rangle_z \triangleq \begin{pmatrix} 1 \\ 0 \end{pmatrix} \quad \text{and} \quad |-\hbar/2\rangle_z \triangleq \begin{pmatrix} 0 \\ 1 \end{pmatrix} \quad (10.3)$$

We can check this explicitly:

$$\hat{S}_z |\hbar/2\rangle_z = \frac{\hbar}{2}\begin{pmatrix} 1 & 0 \\ 0 & -1 \end{pmatrix}\begin{pmatrix} 1 \\ 0 \end{pmatrix}$$

$$= \frac{\hbar}{2}\begin{pmatrix} 1 \\ 0 \end{pmatrix} = \frac{\hbar}{2}|\hbar/2\rangle_z \quad \checkmark$$

and

$$\hat{S}_z |-\hbar/2\rangle_z = \frac{\hbar}{2}\begin{pmatrix} 1 & 0 \\ 0 & -1 \end{pmatrix}\begin{pmatrix} 0 \\ 1 \end{pmatrix}$$

$$= \frac{-\hbar}{2}\begin{pmatrix} 0 \\ 1 \end{pmatrix} = \frac{-\hbar}{2}|-\hbar/2\rangle_z \quad \checkmark$$

As always in Quantum Mechanics, a general state is not a spin-eigenstate, but a superposition of the form $|X\rangle = a\,|\hbar/2\rangle_z + b\,|-\hbar/2\rangle_z$. The coefficients a and b depend on how exactly we prepare our given system.

For example, if we measure the spin along the z-axis and filter out all particles with spin $-\hbar/2$, the coefficient b would be zero and a would be 1. In words this means that after the filtering the probability to measure the value $\hbar/2$ for the spin along the z axis is 100% since we filtered out all particles with the only other allowed value:

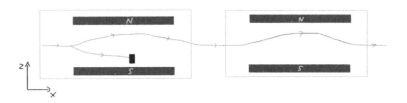

Without any measurement and filtering the coefficients are $a = b = \frac{1}{\sqrt{2}}$, which correspond to the probabilities $|a|^2 = |b|^2 = \frac{1}{2} = 50\%$ for each of the two possibilities.

Now things get *really* interesting if we make a measurement along the z-axis and afterwards a measurement of the spin along a different axis. For concreteness let's say we measure in this second step the spin along the x-axis. Unsurprisingly, even if we filter out all particles with spin down along the z-axis, there will be particles with spin down along the x-axis. However, something cool happens when we measure the spin along the z-axis again after we measured the spin along the x-axis and did a similar filtering procedure here. Even though we filtered out all particles with spin down along the z-axis in the first step, after the filtering regarding the spin in the x-direction, we suddenly get particles again with spin down in the z-direction.

So concretely we perform the following steps

▷ We start with a spin measurement along the z-axis and filter out all particles with spin down ($-\hbar/2$). We describe the resulting state by the following ket

$$|X\rangle_{\text{after z-axis filtering}} = |\hbar/2\rangle_z \, . \quad (10.4)$$

This is a state that purely consists of spin up since we filtered out all particles with spin down. Thus, when we measure the spin along the z-axis after this filtering we get a very unsurprising result: the probability to measure spin $-\hbar/2$ is zero and 100% for spin $+\hbar/2$:[7]

$$_z\langle \hbar/2|X\rangle_{\text{after z-axis filtering}} = 1$$
$$_z\langle -\hbar/2|X\rangle_{\text{after z-axis filtering}} = 0 \, .$$

[7] This is how we extract the probability amplitudes for specific outcomes in the quantum framework: we multiply the ket that describes our state from the left-hand side with the bra that denotes the outcome. We discussed this in Section 3.2.

▷ We then measure the spin along the x-axis. Since we didn't care about the spin in the x-direction in the first step, we get the probabilities $1/2 = 50\%$ for a measurement of spin up in the x direction and a probability of $1/2 = 50\%$ for a measurement of spin down[8].

[8] We will calculate this in a moment.

▷ We then filter out all particles with spin down in the x-direction. This means the particle beam is then in the state

$$|X\rangle_{\text{after x-axis filtering}} = |\hbar/2\rangle_x \, .$$

▷ Finally, we measure the spin in the z-direction again. Here we get the surprising result that the probability for measuring spin down is no longer zero, but $1/2 = 50\%$ instead!

So the bottom line is that the measurement of the spin in the x-direction "produces" spin down states in the z-direction even though we did filter all such states out in the first step. In other words, the measurement of spin in the x direction erases all information we previously collected about spin in the z-direction. This is completely analogous to how a measurement of the momentum changes what we measure for the location[9].

[9] We talked already about this in Section 3.4 and Section 3.10

We now calculate the probabilities explicitly that we already listed in the summary above.

But before we start a few general remarks and reminders:

To find the probability to get the result $-\hbar/2$ for a spin measurement along the z-axis of a system described by a general ket $|X\rangle = a\,|\hbar/2\rangle_z + b\,|-\hbar/2\rangle_z$, we have to calculate[10]

$$_z\langle -\hbar/2|X\rangle = a\,\underbrace{_z\langle -\hbar/2|\hbar/2\rangle_z}_{=0} + b\,\underbrace{_z\langle -\hbar/2|-\hbar/2\rangle_z}_{=1} = b, \quad (10.5)$$

[10] The subscripts always denote regarding which axis a given state is an eigenstate.

where we used that the states $|\hbar/2\rangle$ and $|-\hbar/2\rangle$ are orthogonal and normalized. The probability for measuring $-\hbar/2$ for the spin along the z-axis is then $P_{z=-\hbar/2} = |b|^2$. Now if we want to measure the spin along another axis, say the x-axis, we first need to expand our state in terms of the eigenstates of $\hat{S}_x$. This operator reads in explicit matrix form (Eq (3.53))

$$S_x = \begin{pmatrix} 0 & \hbar/2 \\ \hbar/2 & 0 \end{pmatrix}. \quad (10.6)$$

The corresponding normalized eigenvectors are

$$|\hbar/2\rangle_x \triangleq \frac{1}{\sqrt{2}}\begin{pmatrix} 1 \\ 1 \end{pmatrix} \quad \text{and} \quad |-\hbar/2\rangle_x \triangleq \frac{1}{\sqrt{2}}\begin{pmatrix} 1 \\ -1 \end{pmatrix}. \quad (10.7)$$

So if we want to calculate the probability to measure $-\hbar/2$ for the spin in the x-direction, we first need to rewrite $|\hbar/2\rangle_z$ and $|-\hbar/2\rangle_z$ in terms of $|\hbar/2\rangle_x$ and $|-\hbar/2\rangle_x$:

$$|\hbar/2\rangle_z = \frac{1}{\sqrt{2}}\Big(|\hbar/2\rangle_x + |-\hbar/2\rangle_x\Big)$$

since $\begin{pmatrix} 1 \\ 0 \end{pmatrix} = \frac{1}{\sqrt{2}}\Big(\frac{1}{\sqrt{2}}\begin{pmatrix} 1 \\ 1 \end{pmatrix} + \frac{1}{\sqrt{2}}\begin{pmatrix} 1 \\ -1 \end{pmatrix}\Big)$ (Eq. 10.3 and Eq. 10.7)

and

$$|-\hbar/2\rangle_z = \frac{1}{\sqrt{2}}\Big(|\hbar/2\rangle_x - |-\hbar/2\rangle_x\Big)$$

since $\begin{pmatrix} 0 \\ 1 \end{pmatrix} = \frac{1}{\sqrt{2}}\Big(\frac{1}{\sqrt{2}}\begin{pmatrix} 1 \\ 1 \end{pmatrix} - \frac{1}{\sqrt{2}}\begin{pmatrix} 1 \\ -1 \end{pmatrix}\Big)$ (Eq. 10.3 and Eq. 10.7)

Our general state reads in this new basis[11]

[11] What we do here is a basis change from the basis which is given by the eigenvectors of $\hat{S}_z$ to the basis that is given by the eigenvectors of $\hat{S}_x$. We discussed such a basis change already in Section 3.3.

$$|X\rangle = a\,|\hbar/2\rangle_z + b\,|-\hbar/2\rangle_z$$
$$= a\frac{1}{\sqrt{2}}\Big(|\hbar/2\rangle_x + |-\hbar/2\rangle_x\Big) + b\frac{1}{\sqrt{2}}\Big(|\hbar/2\rangle_x - |-\hbar/2\rangle_x\Big).$$

The probability amplitude to measure $-\hbar/2$ for the spin along the x-axis is therefore[12]

$$_x\langle -\hbar/2|X\rangle = {}_x\langle -\hbar/2|\left(a\frac{1}{\sqrt{2}}\left(|\hbar/2\rangle_x + |-\hbar/2\rangle_x\right)\right.$$
$$\left. + b\frac{1}{\sqrt{2}}\left(|\hbar/2\rangle_x - |-\hbar/2\rangle_x\right)\right)$$
$$= \frac{a}{\sqrt{2}} - \frac{b}{\sqrt{2}}. \qquad (10.8)$$

[12] The probability is
$$P_{x=-\hbar/2} = |\frac{a}{\sqrt{2}} - \frac{b}{\sqrt{2}}|^2$$

After these general remarks, we are finally ready to calculate the probabilities for the example we discussed at the beginning of this section.

We start by filtering out all particles with spin down in the z-direction. The resulting ket is

$$|X\rangle_{\text{after z-axis filtering}} = |\hbar/2\rangle_z. \qquad (10.9)$$

Using our formula from above, we see immediately that the probability to measure $-\hbar/2$ for the spin along the x-axis is[13]

$$P_{x=-\hbar/2} = |\frac{1}{\sqrt{2}} - \frac{0}{\sqrt{2}}|^2 = 1/2.$$

[13] Here we use Eq. (10.8) with $a = 1$ and $b = 0$.

We then filter out all particles with spin $-\hbar/2$ in the x-direction. The resulting ket reads

$$|X\rangle_{\text{after x-axis filtering}} = |\hbar/2\rangle_x. \qquad (10.10)$$

The crucial question is now: what's the probability to measure $-\hbar/2$ for the spin in the z-direction?

To calculate this, we first need to write $|\hbar/2\rangle_x$ in terms of the z-basis states $|\hbar/2\rangle_z$ and $|-\hbar/2\rangle_z$:

$$|X\rangle_{\text{after x-axis filtering}} = |\hbar/2\rangle_x = \frac{1}{\sqrt{2}}\left(|\hbar/2\rangle_z + |-\hbar/2\rangle_z\right)$$

$$\text{since} \quad \frac{1}{\sqrt{2}}\begin{pmatrix}1\\1\end{pmatrix} = \frac{1}{\sqrt{2}}\left(\begin{pmatrix}1\\0\end{pmatrix} + \begin{pmatrix}0\\1\end{pmatrix}\right) \quad \text{(Eq. 10.3 and Eq. 10.7)}$$

The probability to measure $-\hbar/2$ is therefore

$$P_{z=-\hbar/2} = |{}_z\langle -\hbar/2|X\rangle|^2 = 1/2 \neq 0.$$

So, as promised, a measurement of the spin in the x-direction erases all information we previously gathered for the spin in other direction. As a reminder: We filtered out all particles with spin down in the z-direction in the first step. Nevertheless, in the final step, we get a nonzero probability to measure spin down in the z-direction. This is a result of our measurement of the spin in the x-direction since without it the probability to measure spin down in the z-direction would be zero.

Another important aspect of Quantum Mechanics that we haven't talked about so far are systems in which more than one particle with spin are present. In such systems, there are various ways how the individual spins of the particles can add up to the total angular momentum of the system[14]. In the next section we discuss how we can calculate the probabilities to find the system with different values of the total angular momentum. This is important, for example, for the Hydrogen atom. The proton and the electron both have spin $1/2$, and there are various possibilities how they can add up. Depending on their relative alignment the energy of the electron is a bit higher or lower. So a detailed understanding of how spins add up allows us to calculate the energy spectrum of the Hydrogen atom (and lots of other systems) much more precisely.[15]

[14] Think: two vectors that either point in the same or in opposing directions. Adding them yields something large in the first case and zero in the second (if the vectors have the same length).

[15] In Section 7.3 we ignored such effects completely. However, they can be treated as additional corrections and we will talk about how we can include such corrections in Chapter 12.1..

10.2 Spin Addition

For concreteness let's consider two particles with spin $1/2$. We already know that for such spin $1/2$ particles there are always only two possible spin alignments: spin up $|\uparrow\rangle$ or spin down $|\downarrow\rangle$. We are now first interested in the total spin of the system in the z-direction. The possible arrangements of the two spins are as follows:

$$\uparrow\uparrow, \uparrow\downarrow, \downarrow\uparrow, \downarrow\downarrow \, .$$

Here the first arrow represents the first particle (say, the electron) and the second arrow represents the second particle (say, the proton).

What values would we measure if the spins align like this?

The spins in the z-direction simply add up:[16]

$$S_z \psi_1 \psi_2 = \left(S_z^{(1)} + S_z^{(2)}\right) \psi_1 \psi_2 = \left(S_z^{(1)} \psi_1\right) \psi_2 + \psi_1 \left(S_z^{(2)} \psi_2\right)$$
$$= (\hbar m_1 \psi_1) \psi_2 + \psi_1 (\hbar m_2 \psi_2) = \hbar (m_1 + m_2) \psi_1 \psi_2.$$

[16] Reminder: $\hat{S}_z |m\rangle = \hbar m |m\rangle$.

The possible values for the overall spin in the z-direction are therefore

$$\uparrow\uparrow: m = 1$$
$$\uparrow\downarrow: m = 0$$
$$\downarrow\uparrow: m = 0$$
$$\downarrow\downarrow: m = -1$$

The physical interpretation of this observation is that the two spin 1/2 particles can form together a system with total spin 1 or total spin 0. Depending on the alignment of the spin in the case where it is 1, we can measure for the z-components the values −1, 0 or 1. We say the state is in a **triplet state** and denote it by

$$|11\rangle = \uparrow\uparrow$$
$$|10\rangle = \frac{1}{\sqrt{2}} \uparrow\downarrow + \frac{1}{\sqrt{2}} \downarrow\uparrow$$
$$|1-1\rangle = \downarrow\downarrow.$$

On the left-hand side, we have kets that describe the total system. The first number is the total spin of the system and the second number is the z-component. On the right-hand side we can see how the system is constructed in terms of the spins of the individual particles. The factor $\frac{1}{\sqrt{2}}$ is a normalization constant that makes sure that the overall probability is exactly 1.

In words this means that if the complete system consisting of the two particles is in the $|10\rangle$ state, there is a 50% chance to find the spins aligned like this: ↑↓ (first particle spin up, second one spin down), and a 50% chance to find them like this: ↓↑ (first particle spin down, second one spin up). The other possibility is that the complete system has a total spin of zero. In this case, we say the system is in a **singlet state**. In terms of the individual spins this state reads

$$|00\rangle = \frac{1}{\sqrt{2}} \uparrow\downarrow - \frac{1}{\sqrt{2}} \downarrow\uparrow .$$

So the only difference to the $|10\rangle$ state is here a minus sign, which however is crucial if we determine the energy levels, for example, in the Hydrogen atom.

We will not discuss how these factors can be calculated in detail. These numbers $\frac{1}{\sqrt{2}}, -\frac{1}{\sqrt{2}}$ etc. are known as **Clebsch-Gordan coefficients**. They tell us how a system looks like in terms of the individual angular momentums that it consists of.[17] There are clever and general methods to calculate Clebsch-Gordan coefficients. However, for almost any system you can imagine someone has already calculated them. So, you can simply look them up in a table.

[17] The discussion from above works completely analogous if we consider, for example, the addition of an orbital angular momentum and the spin of a particle.

11

Further Systems

There is, of course, an infinite number of quantum systems we did not discuss here. Most of them do not teach us anything new and contain the same lessons hidden behind more complicated mathematics. However, some of the additional systems are so famous that you should know about them nevertheless. In addition, some examples are extremely useful to understand more advanced concepts in a simplified setup. So before we move on to the next part where we discuss clever methods to deal with more complicated systems, a few short comments on other important quantum systems that we do not discuss in detail.

Before we start a short comment on the general pattern:

The main task is always that we have to solve the Schödinger equation for different potentials. For each different system (= different potential) our task is therefore to solve a different **differential equation**. This task is often far from trivial and nothing students can do in an hour or so. In fact, mathematicians are much better at solving differential equations and most of the time we physicists simply use their solutions. You can see that the solutions of the Schrödinger equation for a system are far from trivial if they are named after some person or have at least a special name. Examples we already encountered are the spherical harmonics, the Hermite polynoms, and the Laguerre polynoms. So as soon as we have written down the Schödinger equation for some system, we should always check if it corresponds to a well-known differential equation. Then we can check how the mathematicians solved the problem or simply use their solutions. However, for almost no realistic system such exact solutions exist[1]. The main idea is then to use the solutions of a similar system that can be solved analytically as a starting point and treat the differences between the two systems as perturbations. This is what we will discuss in the next chapter.

[1] The boxes etc. we discussed are merely toy models and nothing we can really observe in nature.

Now, as promised, a few comments on other famous quantum systems and the related buzzwords you should have heard about:

▷ **Three Dimensional Boxes:** While we only discussed one-dimensional boxes, there is nothing really new in two or three dimensions - only a bit more complicated mathematics. A popular example is a particle confined in an infinite three dimensional spherical ball. This means $V = 0$ for $r < a$ and $V = \infty$ otherwise.

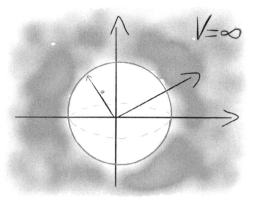

[2] Reminder: for spherically symmetric potentials we can split the variables r and θ, φ and then get two equations: one which only depends on θ, φ and another one that only depends on r. The solutions of the former one are special functions called the spherical harmonics. This is true for all spherically symmetric systems.

As we discussed in Section 7.3, the only thing we have to do here is to solve the radial Schödinger equation since the potential is spherically symmetric[2]. This equation, however, is extremely complicated to solve and the general solutions are another kind of special function called **spherical Bessel functions**. If we then want to determine the energy spectrum of this system we can use that the wave function has to vanish at $r = a$.[3] Therefore we get a condition of the form $\Psi(a, \theta, \varphi) \stackrel{!}{=} 0$. Then we use that the general solutions look like this $\Psi \propto j_n(kr) Y_l^m(\theta, \varphi)$, where $j_n(kr)$ are the spherical Bessel functions and $Y_l^m(\theta, \varphi)$ the spherical harmonics. Our condition $\Psi(a, \theta, \varphi) \stackrel{!}{=} 0$ then tells us immediately that $j_n(kr)$ has to vanish at $r = a$: $j_n(ka) \stackrel{!}{=} 0$. Then, all we have to do is to look up where the zeroes of the spherical Bessel functions are and use that $k \equiv \sqrt{2mE/\hbar^2}$. For example, for $n = 1$ one zero is at $x \approx 4.49$. This tells us that one energy level is at $E = \hbar^2/(2ma^2) \cdot 4.49^2$ since

[3] This is completely analogous to what we did for the one dimensional box. The potential is infinite for $r > a$. Therefore, the wave function is zero in this region. The non-zero wave function inside the box has to connect smoothly to the zero-solution on the outside.

$$j_1(ka) \stackrel{!}{=} 0 \quad \text{and} \quad j_1(4.49) = 0$$
$$\Rightarrow \quad ka \stackrel{!}{=} 4.49$$
$$\sqrt{\frac{2mE}{\hbar^2}} a \stackrel{!}{=} 4.49$$
$$\Rightarrow \quad E \stackrel{!}{=} \frac{\hbar^2}{2ma^2} \cdot 4.49^2 .$$

All other allowed energy values correspond to other zeroes of the spherical Bessel functions. The additional indices m and l in the wave function denote the total angular momentum and the z component of the angular momentum[4] of the particle described by $\Psi \propto j_n(kr) Y_l^m(\theta, \varphi)$. So in summary, we get a quantized energy spectrum again as a result of our boundary condition at $r = a$, analogous to the one dimensional case.

[4] See Section 3.9 and Section 7.3.

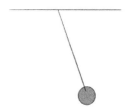

▷ **The Quantum Pendulum**: In the limit where we only consider small displacements the pendulum is very similar to the harmonic oscillator. However, even at low energies, there is something a quantum pendulum can do that a harmonic oscillator cannot. A pendulum can perform a rotation around its revelation. Classically a lot of energy is necessary for such a rotation. However, in Quantum Mechanics the pendulum can tunnel through this potential barrier. An important aspect is that after such a full rotation the pendulum is not necessarily in the same state where it started. Instead, it can pick up a phase. This phase can become physically important, and this makes the pendulum an ideal toy model to understand, for example, the famous quantum phase $\bar{\theta}$ that characterizes the ground state of Quantum Chromodynamics[5]. In addition, the pendulum is a nice toy model to understand why the perturbation series[6] in Quantum Field Theory yields infinity if we add all terms together, but yields sensible results if we only take the first few terms.[7] While the harmonic oscillator is a nice toy model to understand quantum fields with zero spin[8], the pendulum is a perfect toy model to understand quantum fields with nonzero spin[9]. The formula for the potential energy of a pendulum is $V \propto (1 - \cos\phi)$ and the corresponding stationary Schrödinger equation (Eq. (6.5)) therefore reads

$$-\frac{\hbar^2}{2m}\frac{\partial^2\psi(\phi)}{\partial\phi^2} + V(x)\psi = E\psi(\phi)$$

$$-\frac{\hbar^2}{2m}\frac{\partial^2\psi(\phi)}{\partial\phi^2} - mgl(1-\cos\phi)\psi(\phi) = E\psi(\phi). \quad (11.1)$$

[5] Quantum Chromodynamics (QCD) is the correct Quantum Field Theory that describes strong interactions, e.g., how quarks interact with each other.

[6] Perturbation theory is the topic of the next few sections.

[7] Mathematically, a series with properties is known as an **asymptotic series**. To learn more about this try http://jakobschwichtenberg.com/divergence-perturbation-series-qft/.

[8] A quantum field with spin zero, in some sense, is an infinite set of coupled harmonic oscillators.

[9] For example, one of the most famous toy models in Quantum Field Theory - called Sine-Gordon model - describes a bunch of coupled pendulums.

When we consider the pendulum system, the relevant variable is the angle ϕ which only takes on values from 0 to 2π. However, it is also possible to consider the same Schrödinger equation for a general x which takes on any value from $-\infty$ to ∞. In this case the Schrödinger equation describes a particle in a periodic potential[10]. The Schrödinger equation for the pendulum is extremely difficult to solve because of the $\cos(\phi)$ term. In mathematics there is a completely analogous differential equation which is known as **Mathieu's differential equation**. As always, the fact that this equation has a special name already indicates that it is far from trivial. The special functions that solve this equation are known as **Mathieu functions**. Therefore, also the system looks quite simple it is non-trivial to solve in Quantum Mechanics.

[10] This system is incredibly important since it is analogous to the situation which we have for an electron in a crystal. The crystal nuclei are much bigger than the electrons and form a lattice. Therefore, they provide a periodic static background potential for the electrons in the crystal. The wave functions in this context are known as **Bloch waves**.

▷ **The driven Harmonic Oscillator**: this system is just a spring with a motor attached to it. It is useful to understand the interplay between quantum theories and gravity. Since the correct theory of Quantum Gravity is not known yet, we have to use clever tricks to get some insights how quantum theories and gravity could fit together. The driven harmonic oscillator is a nice analogue to a quantum theory with an external source, like the gravitational field, which is still treated classically. Using this toy model it is possible to understand, for example, how in the Quantum Field Theory vacuum, particles can be produced if a gravitational field is present. Famous examples of this effect are the Hawking radiation and the Unruh effect.

12

When the Going Gets Tough, the Tough Lower Their Standards

The title of this chapter is actually a quote from a brilliant book by Sanjoy Mahajan[1]. In it, it is followed by the words: "Approximate first, and worry later". This is exactly what this chapter is about.

[1] Sanjoy Mahajan. *The art of insight in science and engineering : mastering complexity*. The MIT Press, Cambridge, Massachusetts, 2014. ISBN 978-0262526548

The main idea here is that often it is absolutely essential that we get rid of our desire to know everything *exactly*. As already indicated at the end of the last part, often it is simply impossible to find an exact solution; especially for any kind of realistic system that we can actually observe in nature. To be able to say anything at all about such systems we need to lower our standards. The best we can do is find approximately correct answers.

This is not really problematic since in practice we are never able to measure anything exactly anyway. So as long as the error we introduce through our approximations is smaller than the experimental errors, everything is fine. In addition, the magnitude of the approximation errors usually depends on how much time we are willing to spent. So with more effort we can - if we want - calculate the results with higher accuracy.

There is a huge toolbox that we can use to calculate approximately correct answers[2]. We will once more focus on the basic ideas behind approximation methods and only comment on more advanced methods.

[2] There are, in fact, whole books on this topic. So not only on perturbation theory but also whole books on perturbation theory in Quantum Mechanics!

12.1 Perturbation Theory

While above I indicated that approximation methods are especially useful for realistic systems, we will once more stick to simpler toy models. The thing is that in more realistic systems the core ideas are too hidden behind complicated formulas.

What we try to do now is to calculate corrections to previous results that arise when our systems become a bit more realistic. For example, we want to calculate the energy levels for an infinite box when the potential inside the box is not everywhere the same. Another example could be if the potential of our system only looks approximately like the potential of a harmonic oscillator. This is, for example, the case for a pendulum. As long as the pendulum swings a little the harmonic oscillator potential is a great approximation. However, upon closer inspection, the pendulum potential and the harmonic oscillator potential do not agree exactly. We can see this mathematically by Taylor expanding the pendulum potential[3]

[3] We use here the Taylor expansion: $\cos(x) \approx 1 - \frac{x^2}{2} + \frac{x^4}{4!} - \ldots$. See Section A.

$$V \propto 1 - \cos\phi \approx 1 - \left(1 - \frac{\phi^2}{2} + \frac{\phi^4}{4!} - \ldots\right) = \frac{\phi^2}{2} - \frac{\phi^4}{4!} + \ldots. \tag{12.1}$$

For small displacements of the pendulum ϕ higher order terms are smaller than low order terms.[4] So if we don't care about details $V \propto \frac{\phi^2}{2}$ is a good approximation. This is exactly the potential of a harmonic oscillator. However, if we want to calculate, say, the energy levels more exactly we should, at least, take

[4] Consider, for example $\phi = 0.1$. Then we have $\phi^2 = 0.01$, $\phi^4 = 0.0001$, $\phi^6 = 0.000001$. This observation tells us that we do not make a huge mistake if we ignore higher order terms.

the term $\propto \frac{\phi^4}{4!}$ into account, too. So the resulting Schrödinger equation reads

$$-\frac{\hbar^2}{2m}\frac{\partial^2 \psi(\phi)}{\partial \phi^2} - mgl(1-\cos\phi)\psi(\phi) = E\psi(\phi)$$

$$\therefore \quad -\frac{\hbar^2}{2m}\frac{\partial^2 \psi(\phi)}{\partial \phi^2} - mgl\left(\frac{\phi^2}{2} - \frac{\phi^4}{4!}\right)\psi(\phi) \approx E\psi(\phi) \quad \text{(using Eq. 12.1)}$$

which we write as $\quad (H_0 + V_1)\psi(\phi) \approx E\psi(\phi),\quad$ (12.2)

where we have *defined* the unperturbed Hamiltonian

$$H_0 \equiv -\frac{\hbar^2}{2m}\frac{\partial^2 \psi(\phi)}{\partial \phi^2} - mgl\frac{\phi^2}{2}\psi(\phi)$$

and the perturbation

$$V_1 \equiv mgl\frac{\phi^4}{4!}\psi(\phi).$$

The system described by the Schrödinger equation in Eq. (12.2) is known as the **anharmonic oscillator**. The thing is now that H_0 is exactly the Hamiltonian of the harmonic oscillator and if we ignore V_1 for a moment, we therefore already know the exact solutions to this problem.

Therefore, our task is to use this knowledge to calculate the energy levels and wave functions of the anharmonic oscillator, i.e. when we additionally take the new term V_1 into account.

Formulated differently, we now want to do the following: Given the energy levels $E_n^{(0)}$ that correspond to the unperturbed Hamiltonian H_0, what are the eigenvalues if we take a perturbation like V_1 into account?

There is a general method to calculate the corrections to the unperturbed eigenvalues and this is the topic of the next section.

12.1.1 General Perturbation Formulas

In general, the situation we are now interested in looks like this

$$H = H_0 + \lambda V, \quad (12.3)$$

where H is the Hamiltonian for the full system, H_0 the Hamiltonian of a system that we have already solved and V is the difference between the two. Here λ is a parameter that we introduce to keep track of the order of perturbation theory.[5] The main idea is that everything we do here only makes sense if V is in some sense small relative to H_0. Concretely this means that our system is very similar to the system described by H_0 and there is only a small difference between H and H_0. In practice, this means that λ is some small parameter that indicates that V is smaller than H_0.

[5] This will make more sense in a moment. The basic idea is that λ helps us to remember how large the error is we are making with our approximations.

Our goal is to calculate the energy levels E_n of this system which correspond to the eigenvalues of H:[6]

[6] This is simply the stationary Schrödinger equation (Eq. (6.5)).

$$H |n\rangle = E_n |n\rangle$$
$$\therefore \quad (H_0 + \lambda V) |n\rangle = E_n |n\rangle \quad \text{(using Eq. 12.3)} \quad (12.4)$$

where $|n\rangle$ are the corresponding eigenstates. This whole procedure only makes sense if we already know the solutions to H_0:

$$H_0 |n\rangle_0 = E_n^{(0)} |n\rangle_0 . \quad (12.5)$$

Our basic task is now to calculate corrections to $|n\rangle^{(0)}$ and $E_n^{(0)}$. Since we use the parameter λ to label the magnitude of the difference between H_0 and V, we also make the following ansatz for the states[7]

[7] This looks somewhat familiar if we recall what we usually do when we calculate a Taylor series.

$$|n\rangle = |n\rangle_0 + \lambda |n\rangle_1 + \lambda^2 |n\rangle_2 + \ldots \quad (12.6)$$

and for the energies

$$E_n = E_n^{(0)} + \lambda E_n^{(1)} + \lambda^2 E_n^{(2)} + \ldots . \quad (12.7)$$

We call $|n\rangle_1$ and $E_1^{(0)}$ the **first order corrections**, $|n\rangle_2$ and $E_2^{(2)}$ the **second order corrections** and so on. Next, we simply put these ansätze into our Schrödinger equation (Eq. 12.4)[8]

[8] For simplicity, we only consider the terms proportional to λ explicitly

$$(H_0 + \lambda V)|n\rangle = E_n|n\rangle$$

$$\underbrace{(H_0 + \lambda V)(|n\rangle_0 + \lambda |n\rangle_1 + \ldots) = E_n(|n\rangle_0 + \lambda |n\rangle_1 + \ldots)}_{\text{Eq. 12.6}}$$

$$\underbrace{(H_0 + \lambda V)(|n\rangle_0 + \lambda |n\rangle_1 + \ldots) = (E_n^{(0)} + \lambda E_n^{(1)} + \ldots)(|n\rangle_0 + \lambda |n\rangle_1 + \ldots)}_{\text{Eq. 12.7}}$$

$$\therefore \quad H_0|n\rangle_0 + \lambda V|n\rangle_0 + \lambda H_0|n\rangle_1 + \lambda^2 V|n\rangle_1 = E_n^{(0)}|n\rangle_0 + \lambda E_n^{(1)}|n\rangle_0 + \lambda E_n^{(0)}|n\rangle_1 + \lambda^2 E_n^{(1)}|n\rangle_1 + \ldots$$

We can now see why the parameter λ is useful. Since by assumption λ is smaller than one we can use it to rank the various terms by their relative importance. The terms without λ are the most important ones. The second most important terms are those that get multiplied by λ, then the terms proportional to λ^2 and so on[9]. So first, we now collect all terms without λ:

$$H_0|n\rangle_0 = E_n^{(0)}|n\rangle_0 \ . \tag{12.8}$$

This is exactly Eq. (12.5) from above.

Then we collect all terms with λ in front of them:[10]

$$\lambda V|n\rangle_0 + \lambda H_0|n\rangle_1 = \lambda E_n^{(1)}|n\rangle_0 + \lambda E_n^{(0)}|n\rangle_1$$

$$\underbrace{V|n\rangle_0 + H_0|n\rangle_1 = E_n^{(1)}|n\rangle_0 + E_n^{(0)}|n\rangle_1}_{\lambda} \ . \tag{12.9}$$

What we now want is an equation for $E_n^{(1)}$ since these are the dominant corrections to the energy levels $E_n^{(0)}$. A clever idea[11] is now to multiply Eq. (12.9) with ${}_0\langle n|$:

$${}_0\langle n|V|n\rangle_0 + {}_0\langle n|\boldsymbol{H_0}|n\rangle_1 = {}_0\langle n|E_n^{(1)}|n\rangle_0 + {}_0\langle n|E_n^{(0)}|n\rangle_1$$

$$\therefore \quad {}_0\langle n|V|n\rangle_0 + \cancel{{}_0\langle n|E_n^{(0)}|n\rangle_1} = {}_0\langle n|E_n^{(1)}|n\rangle_0 + \cancel{{}_0\langle n|E_n^{(0)}|n\rangle_1} \quad \text{(using Eq. 12.8)}$$

$$\therefore \quad {}_0\langle n|V|n\rangle_0 = {}_0\langle n|E_n^{(1)}|n\rangle_0$$

$$\therefore \quad {}_0\langle n|V|n\rangle_0 = E_n^{(1)} \, {}_0\langle n|n\rangle_0 \quad (E_n^{(1)} \text{ is a number, not an operator})$$

$$\therefore \quad {}_0\langle n|V|n\rangle_0 = E_n^{(1)} \quad ({}_0\langle n|n\rangle_0 = 1) \ .$$

So what we are left with is exactly what we want: an equation that tells us how we can calculate $E_n^{(1)}$. It's simply the expectation value of the perturbation potential V with respect to the

[9] For $\lambda < 1$ we have
$$1 < \lambda < \lambda^2 < \ldots$$

[10] We could equally collect those terms with λ^2 in front of them etc. and this would yield formulas for the higher order corrections. If you want to do this, take note that there are additional terms proportional to λ^2 that only appear implicitly in the equations above since we didn't include the $\lambda^2 |n\rangle_2$ and $\lambda^2 E_n^{(2)}$ terms explicitly.

[11] We will see in a moment why this is a clever idea.

unperturbed state $|n\rangle_0$:

$$E_n^{(1)} = {}_0\langle n| V |n\rangle_0 \qquad (12.10)$$

Following completely analogous steps it is possible to derive formulas for $E_n^{(2)}$, $E_n^{(3)}$, ... up to any order we want. For example, the resulting formula for the second order corrections reads

$$E_n^{(2)} = \sum_{m \neq n} \frac{|{}_0\langle m| V |n\rangle_0|^2}{E_n^{(0)} - E_m^{(0)}}. \qquad (12.11)$$

It is clear that the formulas for higher order corrections are increasingly complicated. So we always only calculate as many corrections as we need to compare our calculations to experimental results. Often the first and second order corrections are enough.

Now, before we have a look at a simple example, there is one more thing we need to discuss.

What we did so far is calculate corrections to the energy levels $E_n^{(0)}$. Following similar steps we can also calculate how the kets $|n\rangle_0$ change in the presence of the perturbations[12], i.e., the corrections $|n\rangle_1$, $|n\rangle_2$, etc.

[12] Or alternatively how the wave functions change.

To calculate the first order correction to the unperturbed kets, we start again with Eq. (12.9)[13] and our first step is to rewrite it as follows

[13] The calculation that follows is a bit more involved. So if you're only interested in the basic ideas, you can skip it and jump directly to the example.

$$V |n\rangle_0 + H_0 |n\rangle_1 = E_n^{(1)} |n\rangle_0 + E_n^{(0)} |n\rangle_1$$
$$\therefore \quad (V - E_n^{(1)}) |n\rangle_0 = -(H_0 - E_n^{(0)}) |n\rangle_1. \qquad (12.12)$$

Everything on the left side is known, so this is a differential equation for $|n\rangle_1$ that we have to solve. The crucial idea is now that the unperturbed kets $|n\rangle_0$ are a complete basis. This means we can write *every* possible state in terms of these kets. Therefore, we can also write the perturbed states in terms of the unperturbed states[14]

[14] We discussed basis changes in Section 3.3.

$$|n\rangle_1 = \sum_m c_{nm} |m\rangle_0 . \qquad (12.13)$$

So now our task is to calculate the coefficients c_{nm}.

We put this ansatz into Eq. (12.12)

$$\begin{aligned}(V - E_n^{(1)}) |n\rangle_0 &= -(H_0 - E_n^{(0)})|n\rangle_1 \\ &= -(H_0 - E_n^{(0)}) \sum_m c_{nm} |m\rangle_0 \quad \text{(using Eq. 12.13)} \\ &= -\sum_m c_{nm} H_0 |m\rangle_0 - \sum_m c_{nm} E_n^{(0)} |m\rangle_0 \quad \text{(rearranging)} \\ &= -\sum_m c_{nm} \left(E_m^{(0)} |m\rangle_0 - E_n^{(0)} |m\rangle_0 \right) \quad \text{(using Eq. 12.5)}\end{aligned}$$

To isolate the coefficient c_{nm} we now multiply this equation with ${}_0\langle l|$ and then use that these basis states are orthogonal[15]:

[15] ${}_0\langle l|m\rangle_0 = \delta_{lm}$ where δ_{lm} denotes the Kronecker symbol which is 1 for $l = m$ and zero otherwise.

$$\begin{aligned}{}_0\langle l| (V - E_n^{(1)}) |n\rangle_0 &= -{}_0\langle l| \sum_m c_{nm} (E_m^{(0)} - E_n^{(0)}) |m\rangle_0 \\ &= -\sum_m c_{nm} (E_m^{(0)} - E_n^{(0)}) \, {}_0\langle l|m\rangle_0 \quad (c_{nm}, E_m^{(0)}, E_n^{(0)} \text{ are numbers}) \\ &= \sum_m c_{nm} (E_n^{(0)} - E_m^{(0)}) \delta_{lm} \quad ({}_0\langle l|m\rangle_0 = \delta_{lm}) \\ &= c_{nl} (E_n^{(0)} - E_l^{(0)})\end{aligned}$$

And we can conclude

$$\frac{{}_0\langle l| (V - E_n^{(1)}) |n\rangle_0}{(E_n^{(0)} - E_l^{(0)})} = c_{nl} . \qquad (12.14)$$

We can put this result back into Eq. (12.13) and this yields

$$\boxed{ |n\rangle_1 = \sum_{m \neq n} \frac{{}_0\langle m| (V - E_n^{(1)}) |n\rangle_0}{(E_n^{(0)} - E_m^{(0)})} |m\rangle_0 . } \qquad (12.15)$$

This is the formula that we can use to calculate how the first order perturbations to our unperturbed kets $|m\rangle_0$ looks like.[16]

Now, let's have a look at a simple example.

[16] Take note that we used in our derivation the stationary Schrödinger equation. This means that our equation here is only valid for systems where the potential does not depend on the time. Moreover, the system makes no sense when two unperturbed energy states have the same energy: $E_m^{(0)} = E_n^{(0)}$ for some $m \neq n$. In this case our formula yields infinity. For such systems there are other methods and we will talk about them in Section 12.2.

12.1.2 The Perturbed Infinite Box

Let's imagine there is a perturbation across half of the infinite box that we discussed in Section 7.1.

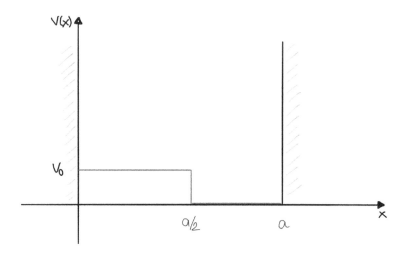

[17] Here we ignore the time-dependent part, i.e., we only focus on the solution of the stationary Schrödinger equation.

The unperturbed energy eigenstates are (Eq. (7.5))[17]

$$\psi_n(x)^{(0)} = \sqrt{\frac{2}{L}} \sin\left(\frac{n\pi}{L}x\right) \qquad (12.16)$$

and the corresponding unperturbed energy levels are (Eq. (7.6))

$$E_n = \frac{n^2 \pi^2 \hbar^2}{L^2 2m}. \qquad (12.17)$$

[18] To get to the second line we use that we can always switch to an explicit basis such as the position basis. The resulting coefficients in this basis are the wave functions $\psi_n(x)^{(0)}$. This was shown explicitly in Eq. (3.27).

Using the general formula (Eq. (12.10)) we can immediately calculate the first order corrections to these energy levels[18]

$$E_n^{(1)} = {}_0\langle n | V | n \rangle_0$$
$$= \int_0^L dx (\psi_n(x)^{(0)})^* V \psi_n(x)^{(0)}$$
$$= \int_0^{L/2} dx (\psi_n(x)^{(0)})^* V_0 \psi_n(x)^{(0)} + \underbrace{\int_{L/2}^L dx (\psi_n(x)^{(0)})^* V \psi_n(x)^{(0)}}_{=0 \text{ since the potential is zero in this second region}}$$
$$= \int_0^{L/2} dx \left(\sqrt{\frac{2}{L}} \sin\left(\frac{n\pi}{L}x\right) \right) V_0 \left(\sqrt{\frac{2}{L}} \sin\left(\frac{n\pi}{L}x\right) \right) \quad \text{(using Eq. 12.16)}$$
$$= \frac{2V_0}{L} \int_0^{L/2} dx \sin^2\left(\frac{n\pi}{L}x\right)$$
$$= \frac{2V_0}{L} \left(\frac{L}{4} - \underbrace{\frac{\sin(Ln\pi/L)}{4n\pi/L}}_{=0} \right) \qquad (\int_0^A \sin^2(xB) = \frac{A}{2} - \frac{\sin(2AB)}{4B})$$
$$= \frac{V_0}{2}.$$

This result does not depend on n and therefore each energy level is simply shifted by the constant amount $\frac{V_0}{2}$ - at least in the first order approximation. This result gives us some confidence in our formula since this is what we would have expected naively.

12.2 What Other Tools Do We Have?

A complete discussion of all existing quantum approximation methods is enough content for at least one complete book. So like in the previous chapters, let's discuss only the main ideas and buzzwords. If you are interested in further details, you can find them in the textbooks recommended in Part 16.

▷ The formulas we derived in Section 12.1.1 are only valid if the Hamiltonian is time-independent since we used in the derivations the stationary Schrödinger equation. There are, of course, also methods to deal with perturbations for systems with a time-dependent Hamiltonian. These tools are known as **time-dependent perturbation theory**. We need these tools, for example, to calculate the rate at which an excited atom returns to its ground state. Formulated differently, to calculate how often a quantum jump happens. Other important applications of time-dependent perturbation theory are scattering processes. The most famous example here is a photon that scatters with an electron or atom. In the latter case, it's possible that the photon gets absorbed and then emitted again. It is possible to calculate the absorption and emission rates and these are especially important in the context of lasers. One main result is **Fermi's golden rule**

$$\Gamma_{i \to f} = \frac{2\pi}{\hbar} |\langle f|V|i\rangle|^2 \rho,$$

which expresses the probability $\Gamma_{i \to f}$ that the initial state i becomes the final state f. Here V is the time-dependent perturbation of the Hamiltonian and ρ the density of final states.

▷ It is also possible that an electron scatters off an atom and no absorption happens. In this case the main goal is to calculate the probability that the electrons gets deflected by a certain angle. The corresponding probability amplitude is usually called **scattering amplitude**. The main approximation methods in this context are known as **partial wave analysis** and the **Born approximation**. One main idea is that we model the atom or whatever we are scattering at by a spherical potential:

$$V(r) = \begin{cases} -V_0 & r \leq a \\ 0 & r > a. \end{cases}$$

The next idea is then that our incoming particle is given by a plane wave and after the scattering we have outgoing spherical waves.

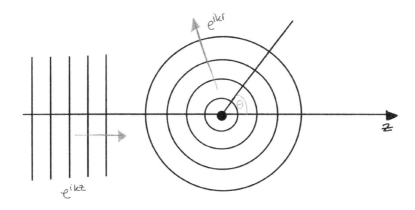

This motivates the ansatz

$$\Psi \approx A \left(e^{ikz} + f(\theta) \frac{e^{ikr}}{r} \right),$$

where e^{ikz} is our incoming plane wave and $\frac{e^{ikr}}{r}$ the outgoing spherical wave. The goal is then to calculate $f(\theta)$ which is the amplitude to find the scattered particle at an angle θ. One main result of the partial wave analysis is that our scattered wave is given by special functions known as **spherical Hankel functions** times the usual spherical harmonics. We expand the general result in terms of these functions and the coefficients C_l in this expansions are known as **partial wave amplitudes**. The amplitude that we want to calculate $f(\theta)$ is directly proportional to these amplitudes:

$$f(\theta) = \frac{1}{k} \sum_{l=0}^{\infty} (-i)^{l+1} \sqrt{\frac{2l+1}{4\pi}} C_l P_l(\cos\theta)$$

where $P_l(\cos\theta)$ are special functions known as **Legendre polynomials**[19]. The main idea that we use to calculate these C_l is once more that inside the spherical potential we have a different wave function than outside of it and that the two kinds of solutions must match up. The Born approximation is a lot more complicated and quickly leads us into the terrain of Quantum Field Theory.

[19] The spherical harmonics are defined in terms of these polynomials.

▷ Another instance when our formulas from Section 12.1.1 fail is if two or more states of the unperturbed system corre-

[20] $|n\rangle_1 = \sum_{m \neq n} \frac{{}_0\langle m|(V-E_n^{(1)})|n\rangle_0}{(E_m^{(0)} - E_n^{(0)})} |m\rangle_0$.

[21] $E_n^{(2)} = \sum_{m \neq n} \frac{|{}_0\langle m|V|n\rangle_0|^2}{E_n^{(0)} - E_m^{(0)}}$.

[22] We shortly talked about the hydrogen atom in Section 7.3.

[23] You can find a full discussion and derivation of the Dirac equation in my book

Jakob Schwichtenberg. *Physics from Symmetry*. Springer, Cham, Switzerland, 2018. ISBN 978-3319666303

spond to the *same* energy level. In such systems, we say that the energy levels are degenerate. We can see that in such systems our formulas fail to produce sensible results by looking at Eq. (12.15)[20] or Eq. (12.11)[21]. If two energies $E_m^{(0)}$ and $E_n^{(0)}$ are equal, we get naively, infinity as a result. The methods that let us make sense of this result and calculate sensible corrections for such systems are known as **degenerate perturbation theory**.

One of the most important applications of degenerate perturbation theory are corrections to the spectrum of the Hydrogen atom.[22] For the naive Hamiltonian we talked about in Section 7.3 there are various states with the same energy. (An exception is the ground state which is non-degenerate.)

How can there be different states with the same energy? Well, these various states with equal energy are characterized by different angular momentum quantum numbers. In physical terms, this means they correspond to different ways how the spin of the electron aligns and adds up with its orbital angular momentum.

Upon closer inspection, we notice that different angular momentum quantum numbers also lead to slightly different potential energies. The **spin-orbit interaction** is described by a new term in the Lagrangian which we can treat as a perturbation to our naive Hamiltonian. The naive Hamiltonian only includes the Coulomb potential. If we take the spin-orbit interaction into account the states with previously the same energy have now different energies and we say the degeneracy has been lifted. Other similarly important corrections come about since we didn't use the correct relativistic energy-momentum relation. The correct relativistic equation to describe electrons is the **Dirac equation** and not the Schrödinger equation.[23]

The Dirac equation takes correctly spin and the relativistic energy-momentum relation into account. We can expand the Dirac equation on a non-relativistic limit and what we end up with is the Schrödinger equation plus higher order correction terms plus a spin term. The spin term describes the spin-orbit interaction we talked about above. The two other equally important correction terms are known as **Darwin term** and **kinetic energy correction**.

Together, these corrections are known as **fine-structure** of the Hydrogen atom. Of course, it is also possible to go beyond that and consider, for example, corrections due to the interaction of the magnetic moment of the electron with the magnetic moment of the proton. The resulting corrections are known as **hyperfine-structure**.

▷ Another handy approximation tool is the so-called **WKB-method**[24]. Here the crucial idea is to make use of the smallness of the Planck constant $\hbar$. To say something is small, of course, only makes sense relative to something else and the thing here is that $\hbar$ is tiny from our macroscopic perspective. So the WKB-method is a **semi-classical approximation method**[25].

This means we treat the system almost as if it were a classical system and then include quantum effects only as corrections. The corrections are given in a power series in $\hbar$, and the whole procedure only makes sense for highly excited states, i.e., large n states[26].

[24] WKB stands for Wentzel-Kramers-Brillouin who popularized the method.

[25] On a first glance, something very confusing is that one of the most important applications of the semi-classical approximation is in the context of tunneling phenomena. For example, we use the semi-classical approximation in Quantum Field Theory to investigate instanton configurations, which are tunneling processes. However, this is possible through a clever trick known as Wick rotation which flips our potential. The behavior in the classically forbidden regions can then be investigated by the semi-classical approximation since through the potential flip these regions become classically allowed.

[26] Recall that n is the quantum number we use to label energies E_n.

Part III
What Your Professor is Not Telling You About Quantum Mechanics

"As with all true and deep physical effects, there are many ways of arriving at the results."

Roman Jackiw

PS: You can discuss the content of Part III with other readers, find exercises to check your understanding and give feedback at www.nononsensebooks.com/qm/part3.

So far we only talked about one possible way to describe what is going on in the quantum world. And most textbooks talk exclusively about this approach. The method we used so far is usually called the wave function formulation.

However, this is not the only possible formulation of Quantum Mechanics and in this final part of the book, we will talk about alternative formulations. The situation is analogous to how we can describe Classical Mechanics either using Newton's formalism, the Lagrangian formalism or the Hamiltonian formalism[27].

[27] There is a one-to-one correspondence between the various formulations of Classical Mechanics and the different formulations of Quantum Mechanics. There is even a formulation of Classical Mechanics that is analogous to the wave function formulation of Quantum Mechanics! We will talk about this in detail in a moment.

Maybe you wonder: why should I care? I already learned one framework that works. So why should I learn anything else?

There are two excellent reasons:

▷ Firstly, it is much easier to describe certain systems in one formulation than in the others. This is analogous to how in Classical Mechanics the generalized coordinates of the Lagrangian formulation often make a problem much simpler than in the Newtonian formulation.

▷ Secondly, knowing the various formulations and how they are connected is extremely useful when you want to think about what Quantum Mechanics really means. In too many discussions people argue about aspects of the wave function and act as if wave functions were something real. If you know that we can describe what is going on in quantum systems completely without wave functions, you will understand that wave functions are merely a convenient mathematical tool.

Before we start, we should pause for a minute and think about what we want to achieve.

First of all, take note that in the following sections, we only talk about different mathematical formulations of Quantum Mechanics and *not* about different interpretations[28]. These different formulations all agree in terms of what they predict

[28] The various interpretations of Quantum Mechanics are the topic of Chapter 15.

for experiments. Only the calculation that we use to get these predictions are different. So none of them is *the* correct one.

The crucial idea that will allow us to describe quantum systems in quite different ways is that there are various mathematical arenas that we can use as the stage where our description of physics takes place. We will discuss various mathematical arenas that are useful in physics in the next chapter. Afterwards, we will see how different the rules of Quantum Mechanics in the various arenas are.

13

Mathematical Arenas

The simplest arena we can use to describe nature is, of course, our everyday space[1]. Here we describe the location and the momentum of each object using an individual vector. These vectors all live in the same arena that we call our everyday space.[2]

For simplicity, let's consider an object which moves in just one dimension. Our mathematical arena is then simply a line ($\mathbb{R}$):

[1] With everyday space I mean the usual Euclidean three-dimensional space $\mathbb{R}^3$ or $\mathbb{R}$ if for some reason our objects can only move in one-dimension.

[2] All this will make a lot more sense as soon as we talk about alternative arenas.

Now, if we want to describe two objects which move in one dimension the first method that comes to our mind is to use two vectors:

In addition to two vectors that keep track of the locations, we need two further vectors that keep track of the momenta.

This is what we do, for example, in the Newtonian formulation of Classical Mechanics. Such a description in everyday space is handy since we can immediately understand everything that is going on in the system. Each vector is simply an arrow that points from one location to another. However, in practice, this approach is often laborious - especially when we are dealing with lots of objects.

So how else could we describe our system consisting of the two objects that move along a line?

What we need, mathematically, is a tool that allows us to keep track of the locations and momenta of the two objects. In the everyday description, we need four vectors to accomplish this: two for the locations and two for the momenta.

Using the following idea, we can describe the whole system with just two vectors[3].

[3] We also need only two vectors if there are three or even more objects in the system.

▷ Firstly, we act as if there were a separate arena for each object:

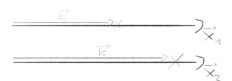

▷ Then we glue these separate spaces together:

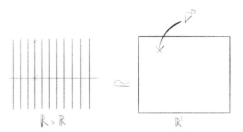

So concretely for the example from above this means that instead of just one line, we now use two. We say our first object moves along one line and the second object along another line. At each possible location of the first object, we need to take into account the possibility that the second object could be anywhere. Hence we need a complete copy of our line $\mathbb{R}$ that we use to keep track of the location of the second object at each location of the $\mathbb{R}$ that we use to keep track the location of the first object. Gluing a copy of $\mathbb{R}$ to each point of $\mathbb{R}$ yields a rectangle[4].

So why is this a clever idea?

Well, instead of two vectors $\vec{r}_1 = (f(x)), \vec{r}_2 = (g(x))$ we can now describe our whole system consisting of the two objects with just one vector $\vec{r} = (f(x), g(x))$. The important point is that this vector lives in a *higher-dimensional* space. So instead of pointing to a point on our line, this new vector $\vec{r}$ points to a point on our rectangle.

In the everyday description, we need N vectors to keep track of the locations of N objects. Using the idea of gluing the spaces together we always only need one vector which, however, lives in an $\mathbb{R}^N$-dimensional space. If the objects are allowed to move freely in three dimensions, our vector $\vec{r}$ lives in $\mathbb{R}^{3N}$ since we are gluing N times $\mathbb{R}^3$ together.

The resulting arena is known as **configuration space**. The basic idea is that instead of keeping track of the N individual objects on our system, we treat the system as a whole. We can imagine the whole system as just one point that moves through this higher-dimensional space called configuration space. Each point in our configuration space corresponds to one specific configuration the system can be in. As time passes on, the configuration of the system usually changes. Using configuration space, we can describe the time evolution of our system by a trajectory in configurations space.

[4] The mathematical name for this kind of construction is **product space**. We will talk about another example of a product space in a moment.

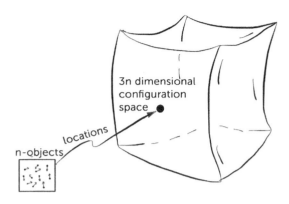

Let's have a look at two concrete examples.

The configuration space of a harmonic oscillator is simply a line[5]

[5] We discussed the quantum harmonic oscillator in Section 9.

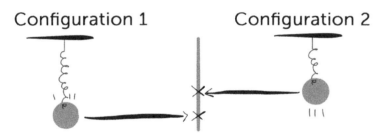

For a second harmonic oscillator our configuration space is also a line, which however we rotate by 90° for reasons that will become clear in a moment:

If we now consider the system that consists of the two harmonic oscillators we need to attach to each point of the configuration space of the first harmonic oscillator the configuration space of the second one. So what we end up with it a rectangle:

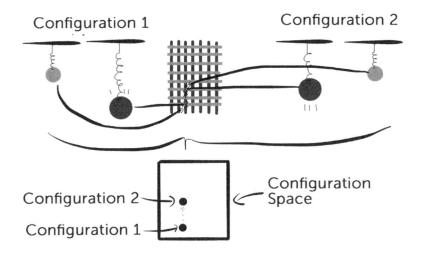

Our second example is a pendulum. The configuration space of a pendulum is a circle since it can rotate around its revelation:

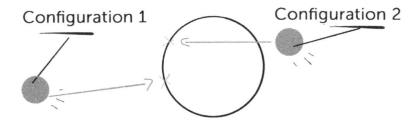

We can then construct the configuration space for the system that consists of two pendulums by attaching to each point of the configuration space of the first pendulum the configuration space of the second one. The result of this procedure is a torus:

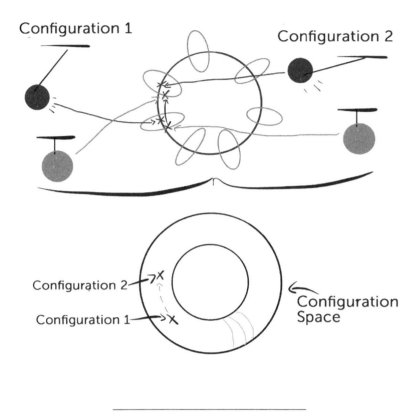

An important observation is that our configuration space only keeps track of the locations of the various objects. However, to describe the state of a system, we need additionally to keep track of their momenta. So in addition to our vector $\vec{r}$ that keeps track of the locations, we need a vector $\vec{p}$ that keeps track of the momenta.

This motivates the construction of the next mathematical arena which works completely analogous to what we did above to construct the configuration space. However, this time we also act as if the momenta live in a different space and then glue the momenta spaces to our location spaces. As a result, we can describe the complete state of our system with just one vector.

The resulting mathematical arena is known as **phase space**. Each point in this phase space corresponds uniquely to one specific location of each object and one specific momentum of each object. So everything that is going on in the system is described by just one vector (or equally the point the vector points to) that moves through phase space.

The price we have to pay for this is that the vector we use to describe the system lives in a $2 \times 3N$-dimensional space for N objects that move in three-dimensions.

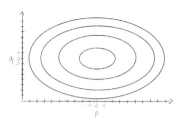

Figure 13.1: Phase space of a harmonic oscillator. The different ellipses correspond to different initial conditions, i.e., different initial velocities and locations of the object at the end of the spring. At the positions where the object is the farthest away from its rest position, the momentum is zero since all kinetic energy is here potential energy saved in the spring. The momentum has its maximum value when the object passes through the rest position since the potential energy is zero here.

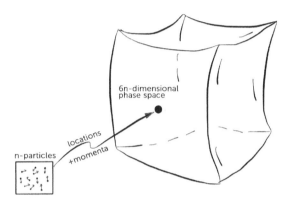

Phase space is notoriously difficult to visualize since even for just two objects moving in one-dimension phase space is already four-dimensional. However, for just one object in one-dimension, it is possible. You can see two examples in Figure 13.2 and Figure 13.2.

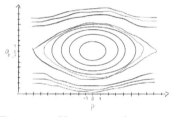

Figure 13.2: Phase space of a pendulum. We can see that for small excitations the system is extremely similar to the harmonic oscillator. However, for large initial momenta, the pendulum can rotate once around its revelation.

So far, we went from our three-dimensional everyday space $\mathbb{R}^3$ via the $3N$-dimensional configuration space to the $2 \times 3N$-dimensional phase space. The final mathematical arena we will discuss next is another development in the same direction. But why do we need another space? Using the corresponding phase space, we can already describe our whole system with just one vector. How could we ever do better than that?

We can't - at least if we were only dealing with situations like the ones described above. However, there is one crucial aspect of every physical system we haven't taken into account yet: uncertainty. In the examples above we assumed that the locations and momenta of every object in the system are perfectly known. In the real world, there is always uncertainty since the precision of a real measurement device is always limited. In addition, as soon as we try to describe the world of elementary particles, there is the quantum uncertainty we seemingly can't get rid of[6]. This feature of nature motivates the construction of a fourth mathematical arena.

[6] We discussed this already in Section 2.3.

Let's recall what uncertainty means. For a single object, it means that we are not completely sure about its momentum and its location. Imagine a beam of particles. We prepare each particle as similar as possible and send them into our experiment separately. However, when we then measure the locations and momenta of different particles we do not always get the same numbers. Instead, our measurement results are always somewhat spread out. This motivates exactly the kind of structure we used all the time in the previous chapter. Instead of a vector which points to one specific location, we now use a superposition of possible locations (or momenta)[7]

$$|\Psi\rangle = a\,|x_1\rangle + b\,|x_2\rangle + c\,|x_3\rangle + d\,|x_4\rangle \,, \qquad (13.1)$$

[7] This is what we already discussed in Chapter 3. However, take note that is also possible to incorporate uncertainty in our descriptions in everyday, configuration and phase space. We will talk about this in detail in the following sections.

where, for simplicity, we here imagine the particles in our system are confined to a lattice structure, i.e., we can only find it at some discrete set of points x_1, x_2, x_3, x_4. The crucial point is now that these objects $|\Psi\rangle$ (our kets) do not live in everyday space or any other space we talked about so far. Instead, they live in a mathematical arena that we usually call **Hilbert space**. And in fact, every calculation in this book did happen in some Hilbert space[8].

[8] For different systems we have different Hilbert spaces, similar to how we have different configuration spaces or different phase spaces for different systems

The main difference to our previous mathematical arenas is that we have *one* basis vector for *each* possible location. So in our example here where only four locations are possible we have four basis vectors: $|x_1\rangle, |x_2\rangle, |x_3\rangle, |x_4\rangle$. However, in general, if we describe an object that moves freely in, say, one dimension we have one basis vector $|x\rangle$ for each point on the line $x \in \mathbb{R}$. Yes, there are infinitely many of them. No matter how much you zoom in, there are always infinitely many points between any two points on the line $\mathbb{R}$.[9] So in general, we now have *in-*

[9] Formulated differently, there are infinitely many real numbers between any given two real numbers.

finitely many basis vectors. Therefore, we can conclude that our mathematical arena is *infinite-dimensional*.

This sounds extremely frightening at first, but luckily we are already somewhat familiar with this kind of structure. We already know how we can squeeze out physical predictions from our infinite-dimensional vectors $|\Psi\rangle$ since this is exactly what we discussed in all previous chapters.

Recall that often we use specific coefficients to describe a given vector $\vec{v} = (3, 1, 5)^T$. Analogously, we often use explicit coefficients $\Psi(x)$ to describe a given vector $|\Psi\rangle$ living in Hilbert space[10]. The only difference is that now we have an expansion with respect to infinitely many basis vectors $|x\rangle$ and we get one coefficient for each point $x \in \mathbb{R}^3$. Therefore, we have infinitely many coefficients $\Psi(x)$ which we call the wave functions. In this sense, people usually say that wave functions live in Hilbert space.

Hilbert spaces are quite similar to the other spaces we discussed previously[11]. The two big differences are that Hilbert spaces are complex and can be infinite-dimensional[12]. By complex we mean that we allow complex linear combinations of the basis vectors, i.e. we allow the coefficients $\Psi(x)$ or a, b, c, d to be complex.[13]

So before we move on and discuss how we can describe quantum systems using the various mathematical arenas discussed in this section, let's summarize what we have learned so far.

▷ One possibility to describe nature is to keep track of everything using vectors living in everyday space.

▷ A bit more convenient is a description in configuration space, where one point is enough to keep track of the locations of all objects in our system.

▷ Even better is a description in phase space, where each point corresponds to one specific state of the system, including all location *and* all momenta.

▷ A fourth possibility is a description in Hilbert space, where

[10] We discussed this in detail in Section 3.3.

[11] Hilbert spaces are also vector spaces, which means that the basic rule to add elements works exactly like for ordinary vectors.

[12] The Hilbert space for specific systems can be finite dimensional. For example, in our example above where only four specific locations were possible, the Hilbert space is finite-dimensional.

[13] We will not discuss the mathematical details of Hilbert spaces any further here. But you can find proper discussions in the textbooks discussed in Part 16.

also each point corresponds to one specific state. However, one advantage is that a Hilbert space is a (complex) vector space. This means that the objects which live in Hilbert space behave exactly like the familiar arrow-type vectors. In contrast, a phase space is, in general, not a vector space[14].

[14] In mathematical terms, we say that phase space is a **symplectic manifold** since the natural product here is the Poisson bracket, which is usually called a symplectic structure.

These are really just different *mathematical* tools that allow us to describe the same systems in different ways. It is up to you which one you like best.

However, the crucial point is the following:

> We can describe any given system
> - classical or quantum -
> using any of these mathematical arenas.

This is how the various formulations of Quantum Mechanics come about and, in fact, also why there are different formulations of Classical Mechanics.

So for Classical Mechanics we have the following more or less well-known formulations:

▷ Classical Mechanics in everyday space is what we call the **Newtonian formulation**.

▷ Classical Mechanics in configuration space is what we call the **Lagrangian formulation**.

▷ Classical Mechanics in phase space is what we call the **Hamiltonian formulation**.

▷ Classical Mechanics in Hilbert space is what we call the **Koopman-von-Neumann formulation**[15].

[15] While the other three are taught in any standard physics curriculum, this fourth formulation is not well-known. For a particularly nice introduction, see the "Notes on Koopman von Neumann Mechanics, and a Step Beyond" by Nobel laureate Frank Wilczek http://frankwilczek.com/2015/koopmanVonNeumann02.pdf.

For Quantum Mechanics, we have completely analogous:

▷ Quantum Mechanics in everyday space is what we call the **pilot wave formulation** or **de Broglie-Bohm formulation**.

▷ Quantum Mechanics in configuration space is what we call the **path integral formulation** or **Feynman formulation**.

▷ Quantum Mechanics in phase space is what we call, well, the **phase space formulation** or **Wigner-Moyal formulation**.

▷ Quantum Mechanics in Hilbert space is what we call the **wave function formulation** or **Schrödinger formulation**[16].

Unfortunately, a full discussion requires at least 100+ pages for each formulation. So the best I can do here is to sketch some of the main ideas and tell you where you can learn more if you are interested. Be warned that I will oversimplify a lot of things in the following sections. My only goal is to give you a rough idea of the main concepts used in the various formulations.

[16] There is a subtlety here since the formulation using bras and kets is often referred to as Dirac formulation. We have already seen above that wave functions only correspond to one specific basis choice. We will talk a bit about another possible formulation in Hilbert space in Section 14.4.

14

The World Beyond Wave Functions

Let's start with the formulation of Quantum Mechanics in the simplest arena that we have: everyday space.

14.1 The Pilot Wave Formulation

I don't know about your preferences, but I think a description of physics in everyday space is potentially extremely useful since we can immediately understand what is going on. While the more abstract configuration, phase, and Hilbert spaces have lots of advantages when it comes to calculations, our everyday space always wins when it comes to intuition.

However, the formulation of Quantum Mechanics in everyday space is - somewhat surprisingly - not very well known and appears in almost no textbook or lecture.

One reason is certainly a historic one. The pilot wave formulation was mainly promoted by the physicists David Bohm and is, in fact, often called Bohmian mechanics. Unfortunately, "*before Bohm could defend his revolutionary ideas about quantum physics, he was swept up in the anti-Communist hysteria of the McCarthy era. Bohm ended up blacklisted and trapped in Brazil, in exile from the rest of the physics world.*"[1]

[1] Adam Becker. *What is real? : the unfinished quest for the meaning of quantum physics.* Basic Books, New York, NY, 2018. ISBN 978-0465096053

When a colleague presented Bohm's alternative formulation at a conference the "*room erupted in vitriol*". And Robert Oppenheimer, then one of the most famous and influential physicists, suggested to the room: "*if we cannot disprove Bohm, then we must agree to ignore him*"[1]. Well, and this is exactly what happened.

So historically Bohm's formulation was dismissed because he was sympathetic towards communism. But why is it still not widely known and taught? His reformulation shouldn't be controversial any longer. It's just a different mathematical way to calculate the same things as in the standard wave function formulation.

I'm not a historian, but my best guess is that Bohm's ideas were dismissed during the crucial years when many important Quantum Mechanics textbooks were written. Since his ideas were dismissed by all leading figures at the time they were not included in any textbook. As a result, the following generations did not learn about his reformulation. When the next generation wrote their textbooks, Bohm's pilot wave formulation again did not get included because the authors either didn't know about it or still thought that there was something wrong with it since so many famous physicists dismissed it. Well, and this story continues to this day. Nowadays, most physicists are simply satisfied with the wave function formulation since it allows them to calculate anything they wish to calculate. Why should they care about another formulation that does not offer any new experimental predictions? This point of view is expressed, for example, by Nobel laureate Steven Weinberg:

In any case, the basic reason for not paying attention to the Bohm approach is not some sort of ideological rigidity, but much simpler - it is just that we are all too busy with our own work to spend time on something that doesn't seem likely to help us make progress with our real problems.[2]

[2] http://inference-review.com/article/on-bohmian-mechanics

So here's the thing. I'm not a hardcore fan of the pilot wave formulation.[3] I don't use it for calculations. I don't think it is in any way superior to the other formulations. But it isn't worse either. So while it makes sense to teach students the wave function formulation, I think, you should at least tell them that such an alternative formulation exists and then let them decide themselves. To me, these different formulations are entirely detached from any ideological arguments about what Quantum Mechanics *really* means[4]. They are simply different *mathematical* tools to calculate the same predictions for what is going on in the quantum world.

[3] There are some die-hard fans of Bohmian mechanics. If you Google for "Bohmian mechanics" you'll find them.

[4] This is the stuff people get passionate about.

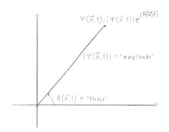

After this long prologue, now let's finally talk about what the pilot wave formulation is all about.

▷ The starting point is the following ansatz for a general wave function:[5]

$$\Psi(\mathbf{r},t) = \sqrt{\rho(\mathbf{r},t)}\, e^{iS(\mathbf{r},t)/\hbar}, \tag{14.1}$$

where $\rho = \Psi\Psi^*$ is the usual probability density[6] and $S(\mathbf{r},t)$ the phase of the wave function.

▷ We put this ansatz into the Schrödinger equation

$$i\hbar\, \partial_t \Psi = \left(-\frac{\hbar^2}{2m}\nabla^2 + V\right)\Psi$$

and then treat the real part and the imaginary part as separate equations. This yields

$$\frac{\partial \rho}{\partial t} + \nabla \cdot \left(\rho \frac{\nabla S}{m}\right) = 0, \tag{14.2}$$

$$\frac{\partial S}{\partial t} + \frac{(\nabla S)^2}{2m} + V + Q = 0, \tag{14.3}$$

[5] We can understand the motivation for this ansatz by observing that wave functions live in a *complex* Hilbert space, but we now want a description in our *real* everyday space. The thing is that we can write any complex number $c \in \mathbb{C}$ as $c = re^{i\theta}$, where r is the absolute value of the complex number and θ its phase. This is known as the **polar form** of the complex numbers and possible thanks to Euler's formula

$$z = e^{ix} = \cos(x) + i\sin(x)$$
$$= \text{Re}(z) + i\text{Im}(z).$$

So instead of the two complex variables c and c^* we can use the two *real valued* variables r and θ to encode the same information.

[6] See Section 3.3

where

$$Q = -\frac{\hbar^2}{8m}\left[2\left(\frac{\nabla^2\rho}{\rho}\right) - \left(\frac{\nabla\rho}{\rho}\right)^2\right]$$
$$= -\frac{\hbar^2}{2m}\left\{\text{Re}\left(\frac{\nabla^2\Psi}{\Psi}\right) + \left[\text{Im}\left(\frac{\nabla\Psi}{\Psi}\right)\right]^2\right\}$$

is known as the **quantum potential**. Equation (14.2) is the usual continuity equation that tells us that probabilities are conserved. The second equation here (Eq. (14.3)) is more interesting since it is completely analogous to the classical Hamilton-Jacobi equation.

▷ The idea is now that we do the same thing as in Classical Mechanics but additionally take the quantum potential into account. As a result of the new potential we get a new force $F_Q = -\vec{\nabla}Q$ that acts on our particles. We can then calculate the trajectories of particles using Newton's usual second law $m\vec{a} = \vec{F} = -\vec{\nabla}(V+Q)$, where V is the normal potential of the system and Q the quantum potential. So that what we end up with here are trajectories in our everyday space, which is exactly what we wanted to accomplish.

▷ The new potential Q is responsible for the strange quantum effects and guides our particles as they move along their trajectories. The thing is that our solutions to Newton's second law $\vec{F} = m\vec{a} = m\ddot{\vec{r}}(t)$ depend crucially on the initial condition $r(t_0)$. However, we *never* know $r(t_0)$ for any particle with absolute accuracy. The quantum potential is extremely sensitive to small changes in the initial conditions. Therefore, the best we can do is average over ensembles of particles. This way we again end up with probabilistic predictions just as in the wave function formulation. The motivation for the name pilot wave formulation comes from the observation that here we have a wave-like potential Q that guides our particles as they move through space. This is in contrast to the wave function formulation where the wave function is a somewhat mysterious object that encodes all information about the system.

It is instructive to see how the famous double slit result comes about in the pilot wave formulation. One crucial observation here is that we never know *exactly* where our particle enters a slit. Have a look at the trajectories for the double slit experiment in the following figure:

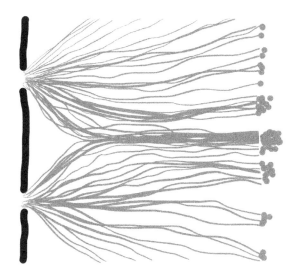

The crucial point is how different the trajectories are for particles that enter the slits right next to each other. This demonstrates how sensitive the quantum potential Q is to the initial conditions. Secondly, the interference pattern comes about because the wave-like quantum potential goes through both slits and then guides the particles in such a way that they end up in the characteristic pattern.

The research group of John Bush recently did a series of illuminating experiments that demonstrated how this can happen.[7] They studied the behavior of oil droplets that bounce along the surface of a liquid. The droplet sloshes the liquid as it bounces along its surface. In return, the path of the droplet gets affected by the ripples in the liquid that are created this way. So the liquid here is the analogue to the quantum potential and the droplet to the quantum particle. Using this setup, they can reproduce several of the most astonishing quantum results,

[7] John Bush is a professor of applied mathematics at the Massachusetts Institute of Technology (http://math.mit.edu/~bush/).

including the result of the double slit experiment[8].

So what's the message to take home here?

Firstly, since all we did was to rewrite the Schrödinger equation, our results are the same as in the standard wave function formulation. And secondly, it's possible - as promised - to describe what is going on in quantum systems using ordinary trajectories in our everyday space.

Two good starting points to learn more about the pilot wave formulation are

▷ **Quantum Theory** by David Bohm[9]

▷ **The Quantum Theory of Motion** by Peter R. Holland[10]

We now move on to the next alternative formulation of Quantum Mechanics which is a lot less controversial. This time our goal is to reformulate Quantum Mechanics in configuration space.

14.2 Path Integrals

We start with a thought experiment that illustrates the general idea[11].

Our starting point is once more the double slit experiment. In the standard wave function formulation, we have a probability amplitude $\psi_1(B)$ that our particle travels from A through slit 1 and then ends up at the location B on our screen. Analogously, we have an amplitude $\psi_2(B)$ that it travels through slit 2 and we then detect it at the location B. The total probability is then the sum of the amplitudes squared[12]

$$P_{AB} = |\psi_{AB}| = |\psi_1(B) + \psi_2(B)|^2. \qquad (14.4)$$

[8] To learn more about these experiments I recommend having a look at the article titled "Fluid Tests Hint at Concrete Quantum Reality" by Natalie Wolchover in Quanta Magazine which is freely available online. In addition, I like the Youtube video "Is This What Quantum Mechanics Looks Like?" by Veritasium https://www.youtube.com/watch?v=WIyTZDHuarQ.

[9] David Bohm. *Quantum theory*. Dover Publications, New York, 1989. ISBN 978-0486659695

[10] Peter Holland. *The quantum theory of motion : an account of the de Broglie-Bohm causal interpretation of quantum mechanics*. Cambridge University Press, Cambridge England New York, NY, 1995. ISBN 978-0521485432

[11] The following thought experiment is due to Anthony Zee and appears in his brilliant book

A Zee. *Quantum field theory in a nutshell*. Princeton University Press, Princeton, N.J, 2010. ISBN 9780691140346

[12] Take note that this is not the same as $|\psi_1(B)|^2 + |\psi_2(B)|^2$. The important difference is the interference term $\psi_1(B)\psi_2(B)$ which is responsible for the interference pattern.

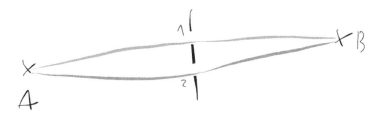

Now, here's a clever series of thoughts which starts with the question: What happens if we drill another slit into our wall?

Well, in this case we simply have

$$P_{AB} = |\psi_{AB}| = |\psi_1(B) + \psi_2(B) + \psi_3(B)|^2. \qquad (14.5)$$

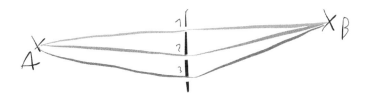

The next clever question is: What happens if we add another wall with holes in it behind the first one?

Again we need to include all possibilities how the particle can get from A to B. This means concretely, for example, that we now have an amplitude for the path from slit 1 in the first wall to the slit $1'$ in the second wall, another amplitude for the path from slit 1 in the first wall to slit $2'$ in the second wall and so on.

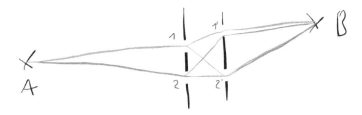

The crazy thing is what all this implies when we take this game to the extreme: We add more and more walls and drill more and more holes into them. At some point there will be no walls left since we drilled so many holes into them. However, our discussion from above suggests how we have to calculate the probability that the particle starts at A and ends up at B: we have to add the amplitudes for all possible paths to get from A to B. This is true even though there are no longer any walls since we drilled so many holes into them. The final lesson is therefore that in empty space without any physical walls, we have to consider the probabilities of the particle taking *all* possible paths from one point to another instead of just one path. This is the basic idea behind the path integral formulation of Quantum Mechanics.

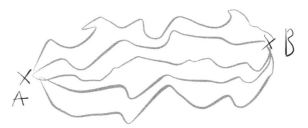

We will now translate exactly this idea into a mathematical form and then see how the name *path integral* comes about.

What we are interested in is the probability that a particle that starts at a point A ends up after some time T at another point B. Using the standard quantum framework, we can immediately write down the corresponding probability amplitude[13]

$$\langle B | \Psi(A,T) \rangle \, . \tag{14.6}$$

[13] We discussed the quantum framework in Chapter 3.

Using the time evolution operator (Eq. (3.44)) we can write this as[14]

$$\langle B | \Psi(A,T) \rangle = \langle B | U(T) | A \rangle$$
$$= \langle B | e^{-iHT} | A \rangle \, .$$

[14] We use here the shorthand notation $|\psi(q)\rangle \equiv |q\rangle$ and, for simplicity assume that the Hamiltonian is time-independent, which is the case for a free particle. Otherwise, we have to write the integral all the time: $U(t) = e^{-\frac{i}{\hbar} \int_0^t dt' H(t')}$. Also, we neglect the factor $\hbar$ here to unclutter the notation.

The thought experiment from above now suggests how we can calculate this: We slice the spatial region between A and B and the time-interval $[0, T]$ into many many pieces. Then to calculate the probability for the particle to go from A to B, we have to sum over the amplitudes for all possible paths between A and B.

For example, let's consider one specific path where the particle travels from A via some intermediate point q_1 to B.

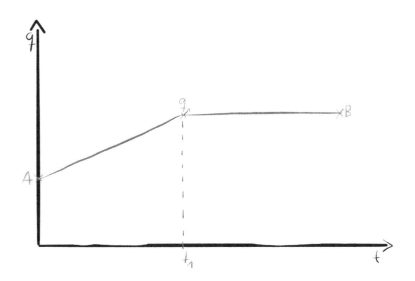

The corresponding probability amplitude is

$$\langle B|e^{-iH(T-t_1)}|q_1\rangle \langle q_1|e^{-iHt_1}|A\rangle ,$$

where t_1 is the time the particle needs to travel from A to the intermediate point q_1.

However, according to our thought experiment, it is not enough to consider one specific path. Instead, we must add the amplitudes for *all* possible paths. This means we need to take into account the probability amplitudes that after t_1 seconds the particle is at *any* possible locations q

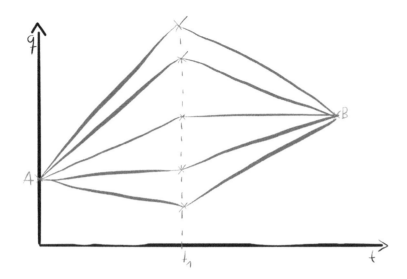

And mathematically this means

$$\sum_{q_1} \langle B|e^{-iH(T-t_1)}|q_1\rangle \langle q_1|e^{-iHt_1}|A\rangle . \tag{14.7}$$

In general, there are not just a discrete set of possible locations after t_1 seconds but instead a continuum. Therefore, we have to replace the sum with an integral

$$\text{Eq. (14.7)} \to \int dq_1 \, \langle B|e^{-iH(T-t_1)/\hbar}|q_1\rangle \langle q_1|e^{-iHt_1}|A\rangle . \tag{14.8}$$

So far we only took into account the probability amplitudes that the particle is at some specific point in time t_1 at all possible locations. However, to consider *all* possible paths we have to do the same thing for all points in time between 0 and T. For this purpose, we slice the interval $[0, T]$ into N equally sized pieces: $\delta = T/N$. The time evolution operator between two points in time is then $U(\delta) = e^{-iH\delta/\hbar}$ and we have to sum after each time evolution step over all possible locations:

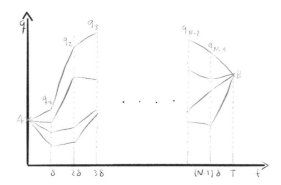

Mathematically, we have completely analogous to Eq. (14.8) for the amplitude $\psi_{A \to B}$ that we want to calculate

$$\psi_{A \to B} = \int dq_1 \cdots dq_{N-1} \langle B| e^{-iH\delta} |q_{N-1}\rangle \langle q_{N-1}| e^{-iH\delta} |q_{N-2}\rangle \cdots$$
$$\cdots \langle q_1| e^{-iH\delta} |A\rangle . \tag{14.9}$$

Our task is therefore now to calculate the products of the form

$$\langle q_{N-1}| e^{-iH\delta} |q_j\rangle \equiv K_{q_{j+1},q_j},$$

which is usually called the **propagator**. We expand the exponential function in a series since δ is tiny[15]

$$K_{q_{j+1},q_j} = \langle q_{j+1}| \left(1 - iH\delta - \frac{1}{2}H^2\delta^2 + \cdots \right) |q_j\rangle$$
$$= \langle q_{j+1}|q_j\rangle - i\delta \langle q_{j+1}| H |q_j\rangle + \ldots . \tag{14.10}$$

[15] Once more we use the Taylor expansion $e^x = \sum_{n=0}^{\infty} \frac{x^n}{n!}$ (Appendix A).

The further evaluation of the propagator is quite complicated and needs many tricks that look extremely fishy at first glance. So don't worry if some steps are not perfectly clear since, as it happens often, you simply need to get used to them.[16] If you're not in the mood for a long calculation, it also makes sense to jump directly to the final result in Eq. (14.15).

[16] The following quote by John von Neumann seems quite fitting here: "Young man, in mathematics you don't understand things. You just get used to them."

With this said, let's continue. The first term in the sum here is a delta distribution since our eigenstates are orthogonal

$$\langle q_{j+1}|q_j\rangle = \delta(q_{j+1} - q_j) = \int \frac{dp_j}{2\pi} e^{ip_j(q_{j+1} - q_j)} . \tag{14.11}$$

[17] This can be motivated as follows: recall that we construct a wave packet as a linear combination of plane waves. The delta distribution is, in a sense, an extreme wave packet which is infinitely thin but at the same time infinitely high. To construct such a wave packet using plane waves we have to use every plane wave that exists. This is basically what we wrote down here. For more on this, see Appendix C.

[18] This is analogous to how we switched from an abstract $|\Psi\rangle$ to the explicit position basis in Section 3.3.

In the last step we rewrote the delta distribution in terms of its explicit integral representation[17].

Next, we evaluate the second term in Eq. (14.10). Here the crucial idea is to recall the explicit form of $H \propto \hat{p}^2/2m + \hat{V}(x)$. To get rid of the operator $\hat{p}$, we need to switch to the momentum basis[18]

$$
\begin{aligned}
-i\delta \langle q_{j+1}| H |q_j\rangle &= -i\delta \langle q_{j+1}| \left(\frac{\hat{p}^2}{2m} + V(\hat{q})\right) |q_j\rangle \\
&= -i\delta \langle q_{j+1}| \left(\frac{\hat{p}^2}{2m} + V(\hat{q})\right) \underbrace{\int \frac{dp_j}{2\pi} |p_j\rangle \langle p_j|}_{=1} |q_j\rangle \\
&= -i\delta \int \frac{dp_j}{2\pi} \left(\frac{p_j^2}{2m} + V(q_{j+1})\right) \langle q_{j+1}| p_j\rangle \langle p_j| q_j\rangle \\
&= -i\delta \int \frac{dp_j}{2\pi} \left(\frac{p_j^2}{2m} + V(q_{j+1})\right) e^{ip_j(q_{j+1}-q_j)},
\end{aligned}
\tag{14.12}
$$

where we again used the orthogonality of the basis states and the explicit integral representation of the delta distribution.

So in summary, our propagator (Eq. (14.10)) now reads

$$
\begin{aligned}
K_{q_{j+1},q_j} &= \langle q_{j+1}| q_j\rangle - i\delta \langle q_{j+1}| H |q_j\rangle + \ldots \\
&\underbrace{=}_{\text{Eq. 14.11}} \int \frac{dp_j}{2\pi} e^{ip_j(q_{j+1}-q_j)} - i\delta \langle q_{j+1}| H |q_j\rangle + \ldots \\
&\underbrace{=}_{\text{Eq. 14.12}} \int \frac{dp_j}{2\pi} e^{ip_j(q_{j+1}-q_j)} \\
&\quad - i\delta \int \frac{dp_j}{2\pi} \left(\frac{p_j^2}{2m} + V(q_{j+1})\right) e^{ip_j(q_{j+1}-q_j)} + \ldots \\
&= \int \frac{dp_j}{2\pi} e^{ip_j(q_{j+1}-q_j)} \underbrace{\left(1 - i\delta \left(\frac{p_j^2}{2m} + V(q_{j+1})\right) + \ldots\right)} \\
&= \int \frac{dp_j}{2\pi} e^{ip_j(q_{j+1}-q_j)} \exp\left(-i\delta \left(\frac{p_j^2}{2m} + V(q_{j+1})\right)\right) \\
&= \int \frac{dp_j}{2\pi} e^{ip_j(q_{j+1}-q_j)} \exp\left(-i\delta H(p_j, q_{j+1})\right)
\end{aligned}
\tag{14.13}
$$

With this at hand, we are finally ready to go back to Eq. (14.9) and evaluate the amplitude $\psi_{A \to B}$. In total, we get N times such a propagator K_{q_{j+1}, q_j}.

$$\psi_{A \to B} = \int \prod_{j=1}^{N-1} dq_j K_{q_{j+1}, q_j}$$

$$= \int \prod_{j=1}^{N-1} dq_j \int \prod_{j=0}^{N-1} \frac{dp_j}{2\pi} \exp\left(i\delta \sum_{j=0}^{N-1} \left(p_j \frac{(q_{j+1} - q_j)}{\delta} - H(p_j, \bar{q}_j)\right)\right) \quad \text{(using Eq. 14.13)}.$$

Now, in the limit $N \to \infty$ our interval δ becomes infinitesimal. Therefore, in this limit the term $\frac{(q_{j+1} - q_j)}{\delta}$ becomes the velocity $\dot{q}$.[19] So the term in the exponent reads $p\dot{q} - H$. If we then execute the integration over dp and recall that the Lagrangian L is exactly the Legendre transform of the Hamiltonian, we can rewrite the amplitude as

[19] This is the definition of the derivative as the difference quotient.

$$\psi_{A \to B} = \left(\frac{m}{2\pi i \delta}\right)^{N/2} \int \prod_{j=1}^{N-1} dq_j \exp\left(i\delta \sum_{j=0}^{N-1} (L(q_j))\right). \quad (14.14)$$

It is conventional to write the amplitude then in the following more compact form[20]

[20] We have included the so-far neglected $\hbar$ again in this final formula.

$$\boxed{\psi_{A \to B} = \int \mathcal{D}q(t) e^{iS[q(t)]/\hbar}} \quad (14.15)$$

where $S[q(t)]$ is the action that we always use in the Lagrangian formalism and $\mathcal{D}q(t)$ the so-called path integral measure.

In words this equation tells us that in Quantum Mechanics we can calculate the probability amplitude that a particle to goes from A to B by summing over all possible paths between A and B and *weight each path by the corresponding action*. This is in stark contrast to what we do in Classical Mechanics. In Classical Mechanics, we also calculate the path an object takes between two fixed points A and B by considering all paths and using the action. But in Classical Mechanics there is only one correct path: the path with minimal action[21].

[21] This is the whole point of the Lagrangian formalism. For each path between two points, we can calculate the corresponding action. Since nature is lazy, she always takes the path with minimal action. The paths with minimal action correspond to solutions of the Euler-Lagrange equations, which are therefore our equations of motion. (In some cases the action is not minimal and a more correct statement would be to talk about paths with extremal action.)

Now, in Quantum Mechanics we act as if the particle takes all possible paths and the classical path with minimal action is therefore only one path out of many.

Take note that the path integral formulation not only works for probabilities that a particle travels between two points. Instead, we can use the same method to calculate the probability that system in a given configuration evolves into another configuration at a later point in time. The path here is a bit more abstract one, namely a path in configuration space.

The explicit evaluation of the path integral (Eq. (14.15)) for concrete systems is notoriously difficult. For almost any system clever approximation schemes are needed to get any information out of it. For this reason, we will not talk about any details here.

Instead, we will discuss a super helpful visual way to understand the path integral which was popularized mainly by Richard Feynman. The main idea is that the action is just a number for any given path. Some paths require a lot of action (i.e., $S[q(t)]$ is large for these paths between A and B) while others require only a little action. The action appears here as the argument of the complex exponential function: $e^{iS[q(t)]}$.

In general, since the action $S[q(t)]$ is an ordinary number this is a complex number with absolute value 1. In the complex plane, these numbers lie on the unit circle[22].

---

[22] Once more we can understand this using Euler's formula

$$z = e^{i\phi}$$
$$= \cos(\phi) + i\sin(\phi)$$
$$= \text{Re}(z) + i\text{Im}(z)$$

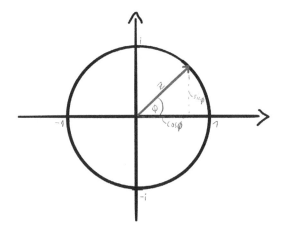

The contribution of each path to the total path integral is therefore simply a unit complex number. The total path integral is a sum over infinitely many unit complex numbers.

Therefore, it is useful to imagine that there is a little stopwatch attached to the particle as it travels a given path. At the beginning of each path the dial points directly to the right[23] which in our complex plane corresponds to $z = 1 = e^{i0}$. Now, as the particle travels the clocks move. At the end of each particular path, the dial points to one specific number on the clock.

[23] On a real clock it would point to the 3.

For example, for one path the situation may look like this:

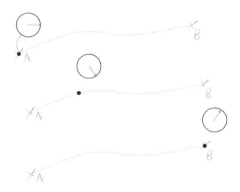

While for another path we have

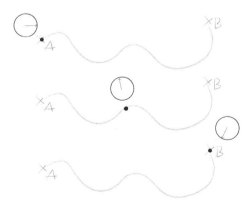

To calculate the path integral, we have to add the little arrows for each path like we would add vectors. The total value of the path integral is then the resulting arrow.

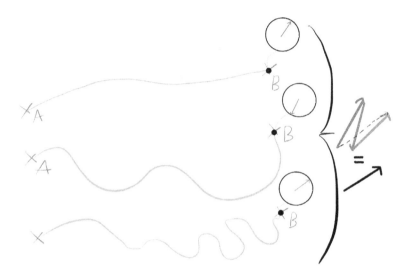

The black arrow here is what we get if we connect the starting point of the blue arrow with the final point of the yellow arrow.

Since the resulting arrows do not necessarily all point in the same direction, the resulting arrow can be quite small. Here, we have three paths but to get the final result we have to include all possible paths, not just three. The final result depends on the starting locations A and B. For some final point B' most of the arrows cancel each other. The resulting arrow is tiny. In physical terms this means the probability to find the particle here is tiny. For another final point B'' lots of arrow point in the same direction and the resulting arrow is large. This means it is quite probable to find the particle here at the end of our time interval.

If you want to learn more about the pictorial way to understand path integrals I highly recommend the book

▷ **QED: The Strange Theory of Light and Matter** by Richard P. Feynman[24].

[24] Richard Feynman. *QED : the strange theory of light and matter.* Princeton University Press, Princeton, NJ, 2014. ISBN 978-0691164090

To learn more about the path integral formulation with real math, a good starting point is

▷ **Quantum Mechanics and Path Integrals** by Richard P. Feynman and A. R. Hibbs[25].

[25] Richard Feynman. *Quantum mechanics and path integrals.* Dover Publications, Mineola, N.Y, 2010. ISBN 978-0486477220

Before we move on there is one more cool thing that we can do with path integrals

14.2.1 The Origin of the Classical Path

First, let's shortly recall what we do in the Lagrangian formulation of Classical Mechanics.

In Classical Mechanics the correct path an object travels is the path with minimal action. A special property of a minimum is that if we investigate its neighborhood, we notice that it only goes up from there and never down. Otherwise, it wouldn't be a minimum. This is the basic idea of **variational calculus**.

To understand this a bit better let's calculate the minimum $x_{\min}$ of some function $f(x) = 3x^2 + x$. We can do this by looking at one specific $x = a$ and then investigate its neighborhood. This means we consider the value of f at $a + \epsilon$, where ϵ denotes an infinitesimal (positive or negative) variation:

$$f(a + \epsilon) = 3(a + \epsilon)^2 + (a + \epsilon) = 3(a^2 + 2a\epsilon + \epsilon^2) + a + \epsilon.$$

Now the crucial idea is that if a is a minimum, the first order variation in ϵ must vanish. Otherwise, we could choose a negative ϵ, and this would mean that $f(a + \epsilon)$ is smaller than $f(a)$. In other words, this requirement makes sure that we don't go upwards if we move away from a. Thus, we now collect all terms linear in ϵ and demand that they vanish:

$$ 3 \cdot 2a\epsilon + \epsilon \stackrel{!}{=} 0 \quad \to \quad 6a + 1 \stackrel{!}{=} 0. $$

From which we can conclude

$$ a \stackrel{!}{=} \frac{-1}{6}. $$

So we have successfully calculated the minimum of our function. This is the same result that we get if use the standard method, i.e., take the derivative $f(x) = 3x^2 + x \to f'(x) = 6x + 1$ and then demand this to be zero[26].

While for functions this is just another method to do the same thing, the variational method is extremely powerful because it also works for functionals[27].

With this in mind we now return to our quantum path integral (Eq. (14.15)). As we will see in a moment, the classical path also plays a special role here, although initially, it seems as if it were just one path out of many.

What we have learned above is that the probability for a given final position depends crucially on the relative positions of the final arrows. If the arrows point mostly in the same direction, we get a long final arrow. We say that in such a situation we have **constructive interference**. If the final arrows point wildly in different directions, they mostly average out, and we end up with a short total arrow. This is known as **destructive interference**.

So why is the classical path (= the path with minimal action) so important?

It turns out that alone it is not that important since every path contributes exactly one arrow. So in some sense, the classical path is in our quantum context just one path out of many. However, we can understand why the classical path is so important in Classical Mechanics by exploring the contributions of neighboring paths. For concreteness, let's consider two neighboring paths $q(t)$ and $q'(t)$ where the second path is a variation of the

[26] This is the standard method to find the minimum of a function. The basic idea here is that the rate of change at a minimum is zero.

[27] A functional is a function of a function. So while an ordinary function eats a number x and spits out a number $f(x)$, a functional $F[f(x)]$ eats a function and spits out a number. An example of a functional is our action $S[q(t)]$ which assigns a number to each path $q(t)$.

first one $q'(t) = q(t) + \eta(t)$, where $\eta(t)$ denotes a small variation.

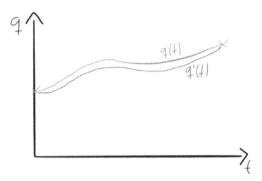

The first path contributes $e^{iS[q(t)]/\hbar}$ while the second path contributes $e^{iS'[q'(t)]/\hbar}$ and we can expand the action of the second path around the first one

$$S[q'] = S[q+\eta] = S[q] + \int dt\, \eta(t) \frac{\delta S[q]}{\delta q(t)} + \mathcal{O}(\eta^2).$$

The thing is now that if $q(t)$ is the path with minimal action $q_{cl}(t)$ the first order variation, as discussed above, vanishes:

$$S[q'] = S[q_{cl}+\eta] = S[q_{cl}] + \underbrace{\int dt\, \eta(t) \frac{\delta S[q]}{\delta q(t)}}_{=0 \text{ for } q(t)=q_{cl}(t)} + \mathcal{O}(\eta^2)$$
$$= S[q_{cl}] + \mathcal{O}(\eta^2).$$

The physical implication for our path integral is now that paths in the neighborhood of the path with minimal action $q_{cl}(t)$ yield arrows that point in approximately the same direction since $S[q'] \approx S[q_{cl}]$. In other words, paths around the classical path interfere constructively. This is why the classical path is important. In contrast, for an arbitrary path far away from the classical path the resulting arrows of neighboring paths vary wildly, and we get destructive interference.

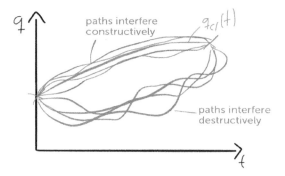

In the next section, we discuss yet another way to calculate probabilities in Quantum Mechanics, this time using the corresponding phase space.

14.3 Phase Space Quantum Mechanics

For a long time people thought that a formulation of Quantum Mechanics in phase space is simply impossible. After all, one of the main lessons of Quantum Mechanics is that we can't measure the location and the momentum of a particle at the same time with arbitrary precision. In a phase space formulation, we use the position and momentum variables at the same time and this seems highly incompatible with Quantum Mechanics[28].

[28] We discussed the general idea behind phase space in Chapter 13.

However, these prejudices are somewhat unfounded since it is quite easily possible to incorporate uncertainty in a phase space formulation. A point in phase space corresponds to one specific state of the system. Formulated differently: such a point corresponds to exactly known locations and momenta for all objects in the system. As time passes by this point moves through phase space since, in general, the locations and momenta of the various objects change. Therefore, a trajectory describes the time evolution of the system in phase space.

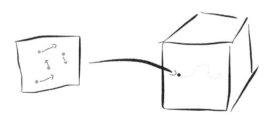

The thing is now that we never know the positions and momenta exactly. There is always some uncertainty. For our phase space, this means that the state of our system does not correspond directly to one point but to a region in phase space. This region consists of all states of the system that are in agreement with what we know about the system.

Maybe the following idea helps: Imagine that if we have definite knowledge about the state of the system we have just one point. If we are not exactly sure this big point splits up into many points which are a bit transparent. The level of transparency indicates how confident we are to find the system in this state. More transparency corresponds to a smaller probability to find the system in the corresponding state. The crucial point is then that if we add up all the points we get a completely non-transparent point which is identical to the definite knowledge point. In physical terms, this means, of course, that probability is conserved. In the definite knowledge case, we have just one point since we are 100% certain that the system is in this state. If we are not sure, the 100% get distributed among all possible states.[29]

[29] We will discuss this using an explicit example in a moment.

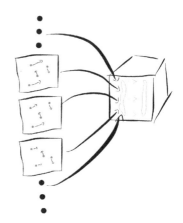

Figure 14.1: In general, for a continuous set of possible initial states, we get a region in phase space.

In practice our limited accuracy means that we don't know the exact point our object is at but only that it has to be in a certain spatial region. The same is true for the momentum which has to be within some range. Now, if we take uncertainty into account, our time evolution is no longer just a trajectory in phase space but a collection of trajectories. Imagine you put a pencil down on each possible initial state of the system. Then, as time passes by each of these pencils traces out the path in phase space that describes the time evolution if the system was in the corresponding initial state. Taken together all of these pencils trace out what we call a **flow** in phase space.

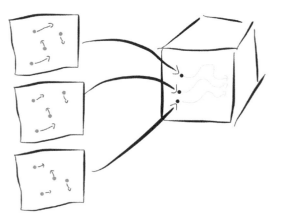

So far, this is nothing that is specifically related to Quantum Mechanics[30]. Everything we just discussed works exactly like

[30] We will discuss what is different in Quantum Mechanics in a moment. But first, we recall how Classical Mechanics works in phase space.

this in Classical Mechanics. The flow of a given distribution of initial states in Classical Mechanics is described by the famous **Liouville equation**

$$\frac{\partial \rho}{\partial t} = -\sum_i \left(\frac{\partial \rho}{\partial q_i} \dot{q}_i + \frac{\partial \rho}{\partial p_i} \dot{p}_i \right)$$
$$= -\{\rho, H\}, \quad (14.16)$$

where $\rho(t, p_i, q_i)$ denotes the **probability density**, $\{\,,\,\}$ the Poisson bracket[31] and H the Hamiltonian function[32].

In words this equation is completely analogous to what we already know from Quantum Mechanics: the Hamiltonian function H generates the time-evolution. The Poisson bracket is the natural product how operators act in phase space. If we integrate this probability density over some phase space volume $p_i \in [p_i^{min}, p_i^{max}]$, $q_i \in [q_i^{min}, q_i^{max}]$, we get the probability to measure the locations and momenta of the objects in our system in the ranges $p_i \in [p_i^{min}, p_i^{max}]$, $q_i \in [q_i^{min}, q_i^{max}]$.

To understand this a bit better, let's imagine a one-dimensional system with two objects and for a moment that for some unspecified reason only very particular positions and momentum values are possible. This means, that only a few initial states are viable and not an infinite continuous set.

Further, let's say we are pretty certain our system is in the state 1 where $q_1 = 2$ m, $p_1 = 3$ kg·m/s and $q_2 = 3$ m, $p_2 = 4$ kg·m/s. However, we can't exclude the state 2 where $q_1 = 3$ m, $p_1 = 4$ kg·m/s and $q_2 = 4$ m, $p_2 = 5$ kg·m/s or state 3 where $q_1 = 1$ m, $p_1 = 2$ kg·m/s and $q_2 = 2$ m, $p_2 = 3$ kg·m/s. Our (now discrete) initial probability density is then[33]

$$\rho(t=0, p_1, q_1, p_2, q_2) = \begin{cases} 0.7, & p_1 = 3, q_1 = 2, p_2 = 4, q_2 = 3 \\ 0.2, & p_1 = 4, q_1 = 3, p_2 = 5, q_2 = 4 \\ 0.1, & p_1 = 2, q_1 = 1, p_2 = 3, q_2 = 2 \end{cases}$$
(14.17)

In words this means that we are 70% certain to find the system it in state 1 and, for example, 20% certain to find it in state 2.

Now, to get the probability at $t = 0$ to measure our objects in

---

[31] The Poisson bracket is defined as follows: $\{F, G\} = \sum_{n=1}^{N} \left(\frac{\partial F}{\partial q_n} \frac{\partial G}{\partial p_n} - \frac{\partial F}{\partial p_n} \frac{\partial G}{\partial q_n} \right)$

[32] The equivalence of the first and second line follows if we use Hamilton's equations:

$$\dot{q} = \frac{\partial H}{\partial p}$$
$$\dot{p} = -\frac{\partial H}{\partial q}.$$

[33] To unclutter the notation we dropped here the units.

the regions $(q_1, q_2) \in \{2\ldots3, 3\ldots4\}$ and with momenta in the range $(p_1, p_2) \in \{3\ldots4, 4\ldots5\}$, we have to "integrate" over the corresponding volume. Since we are dealing with a discrete set of possible states, our "integration" is simply a sum

$$P\Big((q_1,q_2) \in \{2\ldots3,3\ldots4\}, (p_1,p_2) \in \{3\ldots4,4\ldots5\}\Big) = \sum_{q_1=2,q_2=3,p_1=3,p_2=4}^{q_1=3,q_2=4,p_1=4,p_2=5} \rho(0,p_1,q_1,p_2,q_2)$$

$$= \rho(t=0, p_1=3, q_1=2, p_2=4, q_2=3)$$
$$+ \rho(t=0, p_1=4, q_1=3, p_2=5, q_2=4)$$
$$= 0.7 + 0.2 = 0.9 \, .$$

In general, our probability density changes over time, and we can repeat the same procedure for any point in time. The tool that allows us to calculate how any initial probability density evolves in time is Liouville's equation (Eq. (14.16)).

Again: everything we discussed in this section works exactly like this for Classical Mechanics. If we want to understand how we can also describe Quantum Mechanics in phase space, we first need to answer the question:

Where does Liouville's equation come from?

It follows directly from three other equations:

1. **Hamilton's equations**:
$$\dot{q} = \frac{\partial H}{\partial p}$$
$$\dot{p} = -\frac{\partial H}{\partial q} \, .$$

These equations determine the path of each initial state. So in other words, given any state of a system, we can calculate how it evolves in time using Hamilton's equation.

2. The **continuity equation** for the probability density
$$\frac{\partial \rho}{\partial t} = \frac{\partial (\rho \dot{q})}{\partial q} + \frac{\partial (\rho \dot{p})}{\partial p} \, .$$

Using the product rule in this continuity equation and then Hamilton's equations yields exactly Liouville's equation.

The continuity equation tells us that for each initial state we always only get one path through phase space. So in our discrete example from above, if we start with three possible states, at some future point in time, there will again only be three states possible. This follows since each given initial state evolves according to Hamilton's equations. So for each initial state, we get exactly one future state.

Concretely this means that our probability density at some future point in time $t = 10$ s looks like this

$$\rho(t=10, p_1, q_1, p_2, q_2) = \begin{cases} 0.5, & p_1 = 4, q_1 = 22, p_2 = 3, q_2 = 23 \\ 0.4, & p_1 = 3, q_1 = 23, p_2 = 4, q_2 = 24 \\ 0.1 & p_1 = 2, q_1 = 21, p_2 = 5, q_2 = 22 \end{cases}$$

but never like this

$$\rho(t=10, p_1, q_1, p_2, q_2) = \begin{cases} 0.5, & p_1 = 4, q_1 = 22, p_2 = 3, q_2 = 23 \\ 0.2, & p_1 = 3, q_1 = 23, p_2 = 4, q_2 = 24 \\ 0.2, & p_1 = 1, q_1 = 22, p_2 = 1, q_2 = 21 \\ 0.1 & p_1 = 2, q_1 = 21, p_2 = 5, q_2 = 22 \end{cases}$$
(14.18)

This is a crucial assumption. It makes sense in Classical Mechanics since the evolution of each initial configuration is 100% fixed by Hamilton's equations. There may be some uncertainty about which exact initial state is the correct one, but for each possibility, the path through phase space is entirely fixed.[34]

This assumption is often formulated in slogan form: "the phase space flow is incompressible". To understand this imagine that in our discrete example each point in phase space takes up some finite volume. The set of all possible initial configurations then corresponds to some fixed volume in phase space. Now, as time passes on these points move through phase space. However, at some later point in time, if we do the counting, we will again find the same number of points which therefore corresponds to the same phase space volume.

[34] However, there are also systems in Classical Mechanics where this is not true. A famous example is "Norton's dome" or systems with dissipative forces. For these systems, the Liouville equation is wrong since such systems cannot be described using Hamilton's mechanics.

All this talk about volumes in phase space is important in the continuous case where we no longer have a fixed number of points that we can easily follow. So in other words, the crucial assumption that goes into the derivation of Liouville's equation is that the phase space volume that our system takes up in the phase space description always stays the same. However, take note that our phase space density can spread out a lot.

Just imagine how our three discrete points from above can get separated a lot as time passes on. But still, since these three points always stay three points, the total volume they take up is the same. Since a fluid that always takes up the same volume is called incompressible, we borrow this language and usually say that our phase space flow is incompressible[35].

Now, we are finally ready to get back to Quantum Mechanics. Maybe, you can probably already guess how it is different from Classical Mechanics in phase space.

To describe Quantum Mechanics in phase space, we need to get rid of the assumption that our phase space flow is like an incompressible fluid. This is the crucial difference between Quantum Mechanics and Classical Mechanics. And as a result, we have to replace, for example, our Liouville equation with a new equation. But before we talk about these details, let's think about what it means that our phase space flow is no longer incompressible.

To understand this, let's return to our simple example from above. At $t = 0$ we know that the system is in one of just three possible states. Each such state corresponds to one specific point in phase space. If we assume each point takes up some finite unit volume, the total phase space volume our system takes up in phase space is exactly three times this unit volume.

In Classical Mechanics, we are able to trace out the phase space flow by putting down a pencil on each point and then draw the trajectories in phase space for each point as a line. At any later point in time we always end up again with exactly three

---

[35] Think about a bucket of water. We can dump the bucket on the floor. This certainly changes the shape of the water, i.e., the exact form of $\rho(t, p_i, q_i)$. However, the *density* of the water stays the same no matter if it is in the bucket or on the floor. The reason here is of course again that each molecule takes up a fixed amount of space and no molecules get lost if we dump the bucket on the floor.

points and the total phase space volume our description of the system takes up in phase space is constant. The three new points correspond to three new states, and the probabilities can be now very different. But for each initial state, we end up with exactly one final state. Now in Quantum Mechanics if you try to do the same thing with just three pencils, you will fail. The number of possible final states is not necessarily the number of initial states. Or formulated differently, the phase space volume is no longer constant.

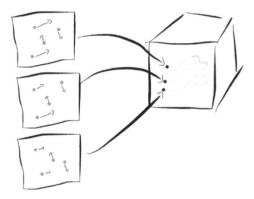

Take note that this does not mean that probability is no longer conserved. This becomes clear by looking at the explicit probability densities for our example above. Initially, there are three possible states (Eq. (14.17)). After some time the probability density possibly looks like the one in Eq. (14.18). While there are now four possible states, the total probability is still 100%. So if we integrate the probability density over all possible states, we always end up with 100%.

Another way how people like to talk about this is in terms of deterministic vs. non-deterministic time evolution[36]. In Classical Mechanics, we have exactly one final state for each initial state. This is a **deterministic evolution**. In contrast, in Quantum Mechanics there can be several possible final states for each initial state. We call this **non-deterministic time evolution**.

To bring this point home here's a quote from one of the most

[36] I'm not a huge fan of these notions since they carry a lot of baggage. There are too many "low signal/high noise" discussions about them. However, still, it's often useful to know the words others like to use.

[37] J. E. Moyal. Quantum mechanics as a statistical theory. *Proc. Cambridge Phil. Soc.*, 45:99–124, 1949. DOI: 10.1017/S0305004100000487

important papers on the phase space formulation of Quantum Mechanics[37]

"Classical statistical mechanics is a 'crypto-deterministic' theory, where each element of the probability distribution of the dynamical variables specifying a given system evolves with time according to deterministic laws of motion; the whole uncertainty is contained in the form of the initial distributions. A theory based on such concepts could not give a satisfactory account of such non-deterministic effects as radioactive decay or spontaneous emission (cf. Whittaker (2)). Classical statistical mechanics is, however, only a special case in the general theory of dynamical statistical (stochastic) processes. In the general case, there is the possibility of 'diffusion' of the probability 'fluid', so that the transformation with time of the probability distribution need not be deterministic in the classical sense."

A crucial observation is that Quantum Mechanics isn't necessarily more uncertain than Classical Mechanics. You can think about the total uncertainty as something like the total distance of the points in phase space.

Say we start with a fairly well-known configuration of the system, which means that all the corresponding points are fairly close to each other in phase space. Now, as time passes on these points can get farther and farther away from each other. And this is in fact what happens in Classical Mechanics. The whole process is known as filamentation of the phase space fluid.

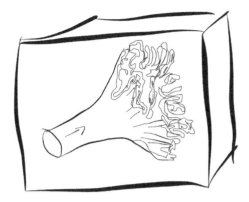

This observation could already be a hint that Classical Mechanics is not the end of the story. To quote Roger Penrose[38]

[38] Roger Penrose. *The emperor's new mind : concerning computers, minds and the laws of physics*. Oxford University Press, Oxford, 2016. ISBN 9780198784920

"For a somewhat analogous situation, think of a small drop of ink placed in a large container of water. Whereas the actual volume of material in the ink remains unchanged, it eventually becomes thinly spread over the entire contents of the container. [...] What this spreading tells us is that, no matter how accurately we know the initial state of a system (within some reasonable limits), the uncertainties will tend to grow in time and our initial information may become almost useless. Classical Mechanics is, in this sense, essentially unpredictable. [...] This spreading effect in phase space has another remarkable implication. It tells us, in effect, that classical mechanics cannot actually be true of our world!"

The reason that Classical Mechanics works fairly well nevertheless is that we usually only consider a small number of large objects where our uncertainties remain manageable.

Now, let's talk about how all this works in practice. I don't want to dive into the mathematical details here, but the main concepts are the following.

The most important difference between Classical Mechanics

[39] Reminder: Liouville's equation reads $\frac{\partial \rho}{\partial t} = -\{\rho, H\}$, where $\{\,,\,\}$ denotes the Poisson bracket: $\{F, G\} = \sum_{n=1}^{N} \left(\frac{\partial F}{\partial q_n} \frac{\partial G}{\partial p_n} - \frac{\partial F}{\partial p_n} \frac{\partial G}{\partial q_n} \right)$.

in phase space and Quantum Mechanics in phase space is that the classical Liouville equation (Eq. (14.16)) gets modified as follows[39]

$$\frac{\partial W}{\partial t} = -\{\{W, H\}\}. \tag{14.19}$$

Here W is the **Wigner quasiprobability distribution** and $\{\{\,,\,\}\}$ denotes the **Moyal bracket**. So in other words, our probability distribution ρ gets replaced with the Wigner quasiprobability distribution and the Poisson bracket with the Moyal bracket. Let's talk about these new objects one after another.

▷ A major new aspect of the Wigner quasiprobability distribution is that it can take on negative values (in extremely localized regions). This feature is also why it's not a probability distribution in the usual sense and thus called a *quasi*probability distribution.

However, the expectation value of some observable A can be calculated using the Wigner function as you would probably expect it

$$\langle A \rangle = \int A(x, p) W(x, p) dx dp. \tag{14.20}$$

In general, we can calculate the Wigner function by using the corresponding wave function $\Psi(x)$[40]

[40] This is the definition of the Wigner quasiprobability distribution.

$$W(x, p) \equiv \frac{1}{\pi \hbar} \int_{-\infty}^{\infty} \Psi^*(x+y) \Psi(x-y) e^{2ipx/\hbar} dy. \tag{14.21}$$

▷ The Moyal bracket in the equation above is defined by

$$\{\{W, H\}\} \equiv -\frac{2}{\hbar} W \sin \left(\frac{\hbar}{2} (\overleftarrow{\partial_x} \overrightarrow{\partial_p} - \overleftarrow{\partial_p} \overrightarrow{\partial_x}) \right) H. \tag{14.22}$$

This is not too illuminating, but if we Taylor expand it powers of $\hbar$, we get

$$\{\{W, H\}\} = \{W, H\} + \mathcal{O}(\hbar^2), \tag{14.23}$$

where $\mathcal{O}(\hbar^2)$ denotes all higher order correction. So the Poisson bracket $\{\,,\,\}$ is the correct approximation for the Moyal bracket for systems where we can ignore corrections proportional to $\hbar$, i.e., quantum corrections. This explains why the Liouville equation works for classical systems.

For more details try

▷ **Quantum Mechanics In Phase Space: An Overview With Selected Papers: An Overview with Selected Papers** by Zachos, Fairlie and Curtright[41]

[41] Cosmas Zachos. *Quantum mechanics in phase space : an overview with selected papers*. World Scientific, New Jersey London, 2005. ISBN 9812383840

Before we summarize what we have learned in the preceding sections, there is one final alternative formulation of Quantum Mechanics we have to talk about. However, this formulation isn't something completely new since it just another formulation in Hilbert space. In some sense, it is only a change of perspective.

14.4 Heisenberg Formulation

In Section 3.6 we discovered that we can describe the time evolution of our states using the Schrödinger equation (Eq. (3.41))

$$i\hbar \partial_t |\Psi(x,t)\rangle = H |\Psi(x,t)\rangle .$$

An important point we didn't talk about so far is that in the wave function formulation we used so far, our operators do not change as time passes on. However, this is not necessarily the case. In fact, we can switch our perspective and reformulate everything such that only the operators change and the states stay the same. In Section 3.6 we also introduced the so-called time evolution operator (Eq. (3.44)):

$$|\Psi(x,t)\rangle = U(t) |\Psi(x,0)\rangle \qquad (14.24)$$

Using this formula and the relationship between bras and kets[42] tells us immediately

[42] Reminder

$$|\Psi(x,t)\rangle^\dagger = \langle \Psi(x,t)|$$

$$\langle \Psi(x,t)| = \langle \Psi(x,0)| U(t)^\dagger . \qquad (14.25)$$

So far, this operator $U(t)$ was just a convenient way to describe the time-evolution of *states* in Quantum Mechanics. However, now comes a crucial idea.

Let's say we want to calculate the expectation value (Eq. (3.18)) of some operator $\hat{O}$ for a system in the state $|\Psi(x,t)\rangle$:

$$\langle \Psi(x,t)|\hat{O}|\Psi(x,t)\rangle = \langle \Psi(x,0)| U^\dagger(t)\hat{O}U(t) |\Psi(x,0)\rangle, \quad (14.26)$$

where we used Eq. 14.24 and Eq. 14.25.

So far, we always assumed that our states change over time. Here this means concretely that our operator $U(t)$ acts on the ket $|\Psi(x,0)\rangle$. However, without changing any result, we can equally say that U acts on the operator $\hat{O}$ instead and the kets remain unchanged. The time evolution of an operator is then given by

$$\hat{O}(t) = U^\dagger(t)\hat{O}U(t). \quad (14.27)$$

Using this, we get the same result for the expectation value:

$$\langle \Psi(x)| \hat{O}(t) |\Psi(x)\rangle = \langle \Psi(x)| U^\dagger(t)\hat{O}U(t) |\Psi(x)\rangle \quad \checkmark$$

This change of perspective is known as the **Heisenberg picture**. In this picture, the operators change while the states remain unchanged. The standard picture we used so far where the states change and the operators remain unchanged is known as the **Schrödinger picture**[43].

In the Heisenberg picture, the Schrödinger equation (which describes the time evolution of states) gets replaced with the so-called **Heisenberg equation**

$$\boxed{\frac{d}{dt}\hat{O} = \frac{i}{\hbar}[\hat{H},\hat{O}] + (\partial_t \hat{O}).}$$

The derivation is completely analogous to the steps in Eq. (4.4) where we derived how Quantum Mechanics is related to Classical Mechanics.

Now it's finally time to summarize what we have learned in the preceding sections.

---

[43] There is a third picture where the operators *and* the states change. Here the idea is similar to what we did in Section 12.1.1 to write the Hamiltonian in two parts $H = H_0 + H_I$ and then split the time-evolution operator

$$U(t) = e^{-\frac{i}{\hbar}\int_0^t dt' H(t')}$$
$$= e^{-\frac{i}{\hbar}\int_0^t dt' (H_0(t')+H_I(t'))}$$
$$= e^{-\frac{i}{\hbar}\int_0^t dt' H_0(t')} e^{-\frac{i}{\hbar}\int_0^t dt' H_I(t')}$$
$$\equiv U_0 U_I.$$

We then say that the operators evolve according to U_0 and the states according to U_I. This third perspective is known as the **interaction picture**. It is especially useful in Quantum Field Theory. The idea is that H_0 describes the free system and H_I all interactions that go on. By using the interaction picture, we can reuse all results for the operators (= the fields) that we derived in the free theory also in the presence of interactions. The only thing that is new is encoded in H_I, and this part only modifies our states.

14.5 Which Formulation Is The Best?

The most important lesson is that no formulation is *the* correct one. All formulations agree in what they predict for experiments. Moreover, in the preceding sections we have explicitly discussed the connections between them. This gives us some confidence in their equivalence.

Nevertheless, each formulation has unique strengths and weaknesses. This is something we should talk about.

▷ The **wave function formulation** has the major advantage that everyone knows and uses it. So if you need any help or want to look up a solution your best chance of finding it is if you use wave functions. In addition, many quantum systems are described most easily using wave functions.

▷ The **pilot wave formulation** is great because it describes quantum systems with ordinary trajectories in everyday space. However, the extreme instability of the quantum potential makes it computationally extremely challenging. Moreover, there are problems to generalize it to Quantum Field Theory.

▷ The **path integral formulation** is extremely useful to investigate formal and global aspects of Quantum Mechanics (and also quantum field theory)[44]. The thing is that since we include *all* paths - even those that move extremely far away and then return to the final location - we grasp global aspects of the system which possibly get lost in other treatments. A huge disadvantage is that the path integral is usually extremely difficult to solve. In addition, there are still problems to make the integral measure $\mathcal{D}q$ that appears in the path integral (Eq. (14.15)) mathematically rigorous.

[44] Famous examples are gauge fixing, Faddeev-Popov ghosts, solitons, instantons and the Fujikawa method to treat anomalies.

▷ The **phase space formulation** makes Quantum Mechanics appear as similar to Hamiltonian mechanics as possible and avoids the operator formalism plus the abstract Hilbert space concept completely. This lets us understand the differences between Classical Mechanics and Quantum Mechanics much more clearly. A significant disadvantage is that this formulation is quite unknown and it's tough to find useful resources.

In the previous sections, we haven't talked about possible interpretations of Quantum Mechanics at all since it is critical to keep the topics "formulations" and "interpretations" separate. But in the following chapter, we will finally talk about what Quantum Mechanics *really* means (or better what people think it could mean).

15

What does it all mean?

"He who confuses reality with his knowledge of reality generates needless artificial mysteries."
Edwin Thompson Jaynes

Before we start, a short disclaimer: I dislike all the fights about what Quantum Mechanics means. However, at the same time, I'm not a fan of the "shut up and calculate" approach either. I think that a proper interpretation of Quantum Mechanics is essential if we want to get a deeper understanding of nature. Moreover, take note that most of what follows is highly subjective and most other physicists have very different views.

With that out of the way, let's dive in.

The most important thing first: Quantum Mechanics *needs* no interpretation. The various formulations of Quantum Mechanics work perfectly without any additional interpretational input. We need no idea about what it all means to describe quantum systems accurately. The equations and algorithms to calculate predictions do not *need* an interpretation to work perfectly.

For this reason Quantum Mechanics, is often taught completely without any discussion about possible interpretations. The whole approach has even its own name: "shut up and calculate". Students are discouraged to "waste" time thinking about what is really going on and should instead focus on learning how to do the calculations.

At the other end of the spectrum are seemingly endless natural language texts on proposed solutions to "problems" and "paradoxes" in Quantum Mechanics. Often it even remains nebulous what the claimed problem is, let alone what the author is proposing. The most important point here is always to keep in mind that all these "problems" people like to argue about are not problems but *puzzles*. We are dealing with a problem when we calculate a prediction that disagrees with what we observe in experiments. And "if quantum theory had been in [such] a crisis, experimenters would have informed us long ago!"[1] In contrast, a puzzle is an aspect of our framework that seems strange or mysterious[2].

So we have "shut up and calculate" on the one end and endless discussions about arcane aspects of the quantum framework on the other end of the spectrum. In-between these two extremes there are lots of intelligent discussions and ideas. Many of the best physicists have joined the debate about what Quantum Mechanics is trying to tell us. Since Quantum Mechanics is now almost a decade old, there are thousands of papers on possible interpretations. So there is no way that we can discuss even a fraction of all the ideas that are out there. For this reason, I will stick to some general comments and list some references where you can learn more.

The most famous interpretation of Quantum Mechanics is the so-called **Copenhagen interpretation**[3]. The goal here is to make sense of the wave function formulation of Quantum Mechanics. Our main focus in this formulation are the measurements we perform on the system described by the corresponding wave function. We have seen that the quantum framework only allows us to make probabilistic predictions.

---

[1] Christopher A. Fuchs and Asher Peres. Quantum theory needs no 'interpretation'. *Physics Today*, 53(3):70–71, March 2000. DOI: 10.1063/1.883004

[2] In fact, most "problems" in modern physics are puzzles. Famous examples are the strong CP "problem", the cosmological constant "problem" or the hierarchy "problem". The crux here is always that some parameter has a value which appears strange and is not what we would naively expect. However, we can simply put in the "strange" experimental values and use our framework without any problems. Of course, it makes sense to wonder about the origin of the strange values, but it is important to keep real problems separate from puzzles.

[3] Something that nicely exemplifies how confusing the whole situation is, is that there isn't even any consensus on what exactly the Copenhagen interpretation is. In fact, while many physicists would answer that it's their preferred interpretation their point of view is often much closer to the "shut up and calculate" approach.

For example, we can only predict that for an ensemble of equally prepared systems, we will measure for 80% of them "spin up", and for 20% of them "spin down". The Copenhagen interpretation states that *through* the measurement process one of the various possibilities gets picked out. Before we measured the spin, the system simply does not have a well-assigned spin. So, in general, any given quantum system does not have definite properties before being measured. The measurement process affects the system in such a way that afterwards, the probability for one of the possibility is 100%. The most famous thought experiment discussed in this context is Schrödinger's cat experiment. Here we put a cat in a box together with a flask of poison and a radioactive source. In the radioactive source, each atom will decay with some probability. When this happens for one atom, it emits a photon which then gets detected by a Geiger counter. This causes a hammer to fall on the flask of poison.

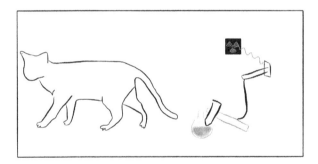

The whole point is now that before we look into the box, we don't know whether the cat is alive or dead. In the Copenhagen interpretation before we open the box, our cat is simultaneously alive and dead since the state reads schematically

$$|\text{cat}\rangle = a\,|\text{cat dead}\rangle + b\,|\text{cat alive}\rangle. \qquad (15.1)$$

However, there are lots of discussions about *when* the measurement here really happens. For example, one can argue that the Geiger counter inside the box already performs the necessary measurement and leads to the collapse of the wave function to $|\text{cat}\rangle = |\text{cat dead}\rangle$. So from this point of view, the collapse does not happen when we open the box but already before that.

Another currently popular interpretation is the so-called **many-worlds interpretation**. Here again the goal is to make sense of the wave function formalism. The basic idea is that the various possibilities encoded in the wave function are all real and realized but not in the same world. For the cat example, this means that as soon as the observer opens the box, the whole system splits into two. In one we have the observer plus the cat alive and in the other one the observer plus a dead cat. At the same time it is impossible that the two newly created "worlds" interact with each other. According to this interpretation, the wave function never collapses. Instead, we have an incredibly large (maybe infinite) number of alternative "worlds" which all correspond to alternative histories and futures.

The Copenhagen interpretation is popular since it was the view held by most of the founding fathers of Quantum Mechanics.[4]. The many-worlds interpretation is fun to think about. However, an incredibly large number of parallel worlds seem like a huge stretch to me. At least as long as there is no experimental hint that at least one such parallel world exists.

[4] It is worth noting that very recently Nobel laureate Steven Weinberg stated in an New York Review of Books essay titled "The Trouble with Quantum Mechanics" that the Copenhagen interpretation "is now widely felt to be unacceptable."

Probably my biggest problem with these interpretations is that they do not take the existence of alternative formulations of Quantum Mechanics into account. They focus solely on the wave function formulation. While wave functions are certainly an extremely convenient tool, there is no need to believe that they are something that exists outside of our description. Instead, there are several hints that wave functions are not real physical entities like, say, the electromagnetic field:[5]

[5] For a very nice discussion along similar lines including comments on how all this came about historically see

Carlo Rovelli. Space is blue and birds fly through it. 2018. DOI: 10.1098/rsta.2017.0312

▷ Probabilities (which we can measure) are only indirectly related to any given wave function: $P = |\psi|^2$. This means we can always multiply our wave function with any complex number of length one: $\psi \to e^{i\phi}\psi$ without changing anything that we can measure

$$P = |\psi|^2 = \psi^*\psi \to \psi^* e^{-i\phi} e^{i\phi} \psi = \psi^*\psi.$$

Any prediction will be the same no matter if we use ψ or $e^{i\phi}\psi$. However, the wave function itself changes dramatically through such a gauge transformation. This is known as **Gauge invariance.**

▷ Wave functions are not objects in our real everyday space but live in abstract Hilbert spaces. So they are not physical entities in the same sense as real waves that we observe in everyday life.

▷ Thirdly, there is the **measurement problem**[6] that arises when we act as if the wave function is a real physical entity. Prior to our measurement of the location of the particle, the corresponding wave function usually spreads out over a large region of space. However, as soon as we measure, the location the wave function is suddenly concentrated to a single point. This is, of course, a non-problem if the wave function is merely a mathematical tool.

[6] Once more we would be better off by calling it the measurement *puzzle*.

The point to take away here is nicely summarized by the following story Felix Bloch told about a conversation he had with Werner Heisenberg:

"We were on a walk and somehow began to talk about space. I had just read Weyl's book Space, Time and Matter, and under its influence was proud to declare that space was simply the field of linear operations. "Nonsense," said Heisenberg, "space is blue and birds fly through it." This may sound naive, but I knew him well enough by that time to fully understand the rebuke. What he meant was that it was dangerous for a physicist to describe Nature in terms of idealised abstractions too far removed from the evidence of actual observation"[7]

[7] Felix Bloch. Heisenberg and the early days of quantum mechanics. *Physics Today*, December 1976

In other words, the main problem is that often physicists have a tendency to identify theoretical constructs (like the wave function) of highly successful models with reality itself. Edwin Jaynes coined the name **mind projection fallacy** for this phenomenon:

This oldest of all devices for dealing with one's ignorance, is the first form of what we have called the "Mind Projection Fallacy". One asserts that the creations of his own imagination are real properties of Nature, and thus in effect projects his own thoughts out onto Nature.

It is still rampant today, not only in fundamentalist religion, but in every field where probability theory is used. [...] [I]n the Copenhagen interpretation of quantum theory, whatever is left undetermined in a pure state ψ is held to be unknown not only to us, but also to Nature herself. That is, one claims that ψ represents a physically real "propensity" to cause events in a statistical sense (a certain proportion of times on the average over many repetitions of an experiment) but denies the existence of physical causes for the individual events below the level of ψ.[8]

[8] E. T. Jaynes. *Probability Theory as Logic*, pages 1–16. Springer Netherlands, Dordrecht, 1990. ISBN 978-94-009-0683-9

With that said, what further alternatives are there?

▷ **The statistical interpretation.** Here the whole idea is that the rules of Quantum Mechanics only apply to ensembles of similarly prepared systems and don't make statements about individual systems. To quote Albert Einstein: "The attempt to conceive the quantum-theoretical description as the complete description of the individual systems leads to unnatural theoretical interpretations, which become immediately unnecessary if one accepts the interpretation that the description refers to ensembles of systems and not to individual systems."[9] To learn more about this interpretation, have a look at "The statistical interpretation of quantum mechanics" by Leslie Ballentine [10].

[9] Paul Schilpp. *Albert Einstein, philosopher-scientist.* Open Court, La Salle, Ill, 1970. ISBN 979-0875482865

[10] L. E. Ballentine. The statistical interpretation of quantum mechanics. *Rev. Mod. Phys.*, 42:358–381, 1970. DOI: 10.1103/RevModPhys.42.358

[11] See, for example, Section 2.3 and Section 2.4 in

James Sethna. *Statistical mechanics : entropy, order parameters, and complexity.* Oxford University Press, Oxford New York, 2006. ISBN 9780198566779

▷ **The stochastic interpretation.** Here the main assumption is that the quantum rules follow since all our particles perform random walks. This is motivated by the observation that the Schrödinger equation is exactly the diffusion equation that describes random walking particles, but with an imaginary diffusion constant[11]. A famous example for random walking particles are pollen grains in water. Their seemingly random behavior is a result of the permanent collisions of the pollen grains with individual water molecules. Since the water molecules are too small to be seen with an ordinary microscope, the pollen grains seem to move randomly.

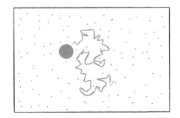

The idea is now that something similar explains the quantum

behavior of elementary particles. Similarly to what happens in the water and pollen grain example, the strange behavior of quantum systems could then be explained once we have better "microscopes" which can reveal the cause for their apparent random motion. One proposal is that vacuum fluctuations and the constant interaction of all particles with them are causing the random walks. To learn more about this approach, have a look at the "Review of stochastic mechanics" by Edward Nelson.[12]

[12] Edward Nelson. Review of stochastic mechanics. *Journal of Physics: Conference Series*, 361 (1):012011, 2012. URL http://stacks.iop.org/1742-6596/361/i=1/a=012011

These interpretations have the advantage that they are not specifically related to one particular formulation of Quantum Mechanics. There are *dozens* of other proposed interpretations. And as already mentioned above there is no way that we can discuss them all. But if you want to learn more about alternative interpretations, there are lots of textbooks to choose from. Here are some good ones:

▷ **Conceptual foundations of quantum mechanics** by Bernard D'Espagnat [13].

[13] Bernard d'Espagnat. *Conceptual foundations of quantum mechanics*. Advanced Book Program, Perseus Books, Reading, Mass, 1999. ISBN 978-0738201047

▷ **The Interpretation of Quantum Mechanics** by Roland Omnes[14].

[14] Roland Omnes. *The interpretation of quantum mechanics*. Princeton University Press, Princeton, N.J, 1994. ISBN 978-0691036694

▷ **Elegance and Enigma - The Quantum Interviews** by Maximilian Schlosshauer[15].

[15] Maximilian Schlosshauer. *Elegance and enigma : the quantum interviews*. Springer, New York, 2011. ISBN 978-3642208799

▷ **Foundations of Quantum Mechanics: An Exploration of the Physical Meaning of Quantum Theory** by Travis Norsen[16]

[16] Travis Norsen. *Foundations of quantum mechanics : an exploration of the physical meaning of quantum theory*. Springer, Cham, Switzerland, 2017. ISBN 978-3319658667

Two final comments before we move on.

If you feel the urge to join the discussion about what Quantum Mechanics means, please do it. We need more smart people with an adequate physical and mathematical background that think deeply about such fundamental issues[17].

In addition, if you want to start thinking about interpretations

[17] Often (young) researchers are actively discouraged to spend time on such problems since many consider it a waste of time. For a nice essay on this topic, see "*Shut up and let me think. Or why you should work on the foundations of quantum mechanics as much as you please*" by Pablo Echenique-Robba; https://arxiv.org/abs/1308.5619

seriously, you should know which often-heard statements about Quantum Mechanics are true and which are not. There are too many myths that are routinely told to students. A good starting point is

▷ **Quantum mechanics: Myths and facts** by Hrvoje Nikolic[18].

[18] Hrvoje Nikolic. Quantum mechanics: Myths and facts. *Found. Phys.*, 37:1563–1611, 2007. DOI: 10.1007/s10701-007-9176-y

16

Get an Understanding of Quantum Mechanics You Can Be Proud Of

As I already warned you in the preface, the content of this book is far from comprehensive. There are hundreds of different aspects of Quantum Mechanics that I didn't even say a word about. Quantum Mechanics is almost a century old now. Thousands of people have worked on it. And unsurprisingly, no single book can capture it all.

However, there are lots of excellent books that cover various special aspects extremely well. There's no need for you to read hundreds of books on Quantum Mechanics - just the best. Since for every good book there are at least 20 bad ones which are not worth your time, below I recommend some of my favorites. So, start by picking the ones that interest you most, and dig in[1].

Two highly recommended books which are written in the same **physics-first spirit** as the book you are currently reading are

[1] If you need further or more specialized reading recommendations, you should visit:
  www.physicstravelguide.com
This is an expository physics wiki where anyone can help to collect the best resources on any physics topic + publish student-friendly explanations.

▷ **Lectures on Physics Volume 3** by Richard Feynman[2].

▷ **Quantum Mechanics** by David J. Griffiths[3].

To get more experience with the quantum framework the following two **problem books** are extremely useful. They contain not only hundreds of problems but also the corresponding solutions.

▷ **Exploring Quantum Mechanics: A Collection of 700+ Solved Problems for Students, Lecturers, and Researchers** by Victor Galitski, Boris Karnakov, Vladimir Kogan, Victor Galitski Jr[4].

▷ **Schaum's Outline of Quantum Mechanics** by Yoav Peleg, Reuven Pnini, Elyahu Zaarur, Eugene Hecht[5]. You can find this book free online; Google it!

To get a better **visual understanding** of how wave functions behave in various systems the following two books are great:

▷ **Visual Quantum Mechanics** by Bernd Thaller[6].

▷ **The Picture Book of Quantum Mechanics** by Siegmund Brandt and Hans Dieter Dahmen[7].

If you're interested in **formal and advanced aspects** of Quantum Mechanics the best starting points are

▷ **Quantum Theory: Concepts and Methods** by Asher Peres[8].

▷ **Quantum Mechanics: A Modern Development** by Leslie E. Ballentine[9].

To understand Quantum Mechanics in more rigorous terms

---

[2] Richard Feynman. *The Feynman lectures on physics*. Basic Books, a member of the Perseus Books Group, New York, 2011. ISBN 978-0465025015

[3] David Griffiths. *Introduction to quantum mechanics*. Pearson Prentice Hall, Upper Saddle River, NJ, 2005. ISBN 9780131118928

[4] V. M. Galitski. *Exploring quantum mechanics : a collection of 700+ solved problems for students, lecturers, and researchers*. Oxford University Press, Oxford, 2013. ISBN 9780199232727

[5] Yoav Peleg. *Quantum mechanics : based on Schaum's outline of theory and problems of quantum mechanics*. McGraw-Hill, New York, 2006. ISBN 9780071455336

[6] Bernd Thaller. *Visual quantum mechanics : selected topics with computer-generated animations of quantum-mechanical phenomena*. Springer/TELOS, New York, 2000. ISBN 9780387989297

[7] Siegmund Brandt. *The picture book of quantum mechanics*. Springer, New York, NY, 2012. ISBN 9781461439509

[8] Asher Peres. *Quantum theory : concepts and methods*. Kluwer Academic, Dordrecht Boston, 1993. ISBN 978-0-7923-2549-9

[9] Leslie Ballentine. *Quantum mechanics : a modern development*. World Scientific, Singapore River Edge, NJ, 2000. ISBN 9789810241056

and learn more about **related mathematical topics**, you should consult the following three books

▷ **Quantum Theory for Mathematicians** by Brian Hall[10]. A free version is available online.

▷ **Quantum Theory, Groups and Representations** by Peter Woit[11]. You can download a free version from Peter Woit's website (www.math.columbia.edu/~woit).

▷ **An Introduction to the Mathematical Structure of Quantum Mechanics** by Franco Strocchi[12].

In addition, to get a feeling for why mathematical rigor can be extremely useful in Quantum Mechanics I recommend the following paper

▷ **Mathematical surprises and Dirac's formalism in quantum mechanics** by Francois Gieres[13].

Finally, there are lots of other good books you should consult whenever you are stuck and confused about something. In such a situation the only way forward is usually to read what different authors have to say about it. Luckily there are dozens of textbooks that cover the standard topics. The following books are all somewhat popular and ideally suited to find a second or third explanation:

▷ **Modern Quantum Mechanics** by Jun John Sakurai[14].

▷ **The Principles of Quantum Mechanics** by Paul Dirac[15].

[10] Brian Hall. *Quantum theory for mathematicians*. Springer, New York, 2013. ISBN 9781489993625

[11] Peter Woit. *Quantum theory, groups and representations : an introduction*. Springer, Cham, Switzerland, 2017. ISBN 9783319646107

[12] F Strocchi. *An introduction to the mathematical structure of quantum mechanics : a short course for mathematicians*. World Scientific, Hackensack, N.J, 2008. ISBN 9789812835222

[13] F. Gieres. Dirac's formalism and mathematical surprises in quantum mechanics. *Rept. Prog. Phys.*, 63: 1893, 2000. DOI: 10.1088/0034-4885/63/12/201

[14] J. J. Sakurai. *Modern quantum mechanics*. Pearson India Education Services, Noida, India, 2014. ISBN 9789332519008

[15] P. A. M. Dirac. *The principles of quantum mechanics*. Clarendon Press, Oxford England, 1981. ISBN 9780198520115

▷ **Quantum Mechanics** by Albert Messiah[16].

▷ **Principles of Quantum Mechanics** by Ramamurti Shankar[17].

▷ **Quantum Mechanics** by Claude Cohen-Tannoudji, Bernard Diu und Franck LaloÃń[18].

The logical next step in the hierarchy of theories after Quantum Mechanics is Quantum Field Theory. Quantum Field Theory is what we end up with if we combine the lessons of Quantum Mechanics with those of Einstein's theory of Special Relativity.

The absolute best book to get started is

▷ **Student Friendly Quantum Field Theory** by Robert D. Klauber[19].

In addition, at this stage, you might also enjoy my other book

▷ **Physics from Symmetry**[20]. Here I argue that the equations used in Quantum Field Theory and all other fundamental theories of modern physics can be understood most easily from a symmetry perspective.

[16] Albert Messiah. *Quantum mechanics*. North-Holland, Amsterdam, 1965. ISBN 9780471597667

[17] Ravi Shankar. *Principles of Quantum Mechanics*. Springer Verlag, City, 2014. ISBN 9781461576754

[18] Claude Tannoudji. *Quantum mechanics*. Wiley, New York, 1977. ISBN 9780471164333

[19] Robert Klauber. *Student friendly quantum field theory : basic principles and quantum electrodynamics*. Sandtrove Press, Fairfield, Iowa, 2014. ISBN 9780984513956

[20] Jakob Schwichtenberg. *Physics from Symmetry*. Springer, Cham, Switzerland, 2018. ISBN 978-3319666303

One Last Thing

It's impossible to overstate how important reviews are for an author. Most book sales, at least for books without a marketing budget, come from people who find books through the recommendations on Amazon. Your review helps Amazon figure out what types of people would like my book and makes sure it's shown in the recommended products.

I'd never ask anyone to rate my book higher than they think it deserves, but if you like my book, please take the time to write a short review and rate it on Amazon. This is the biggest thing you can do to support me as a writer.

Each review has an impact on how many people will read my book and, of course, I'm always happy to learn about what people think about my writing.

So on that note, if you buy my book on Amazon and leave a review, I will send you a free bonus chapter. Simply send me a link or screenshot to your review on Amazon to mail@jakobschwichtenberg.com or via Twitter @JakobSchwich.

Part V
Appendices

A

Taylor Expansion

The Taylor expansion is one the most useful mathematical tools and we need it in physics all the time to simplify complicated systems and equations.

We can understand the basic idea as follows:

Imagine you sit in your car and wonder what your exact location $l(t)$ will be in 10 minutes: $l(t_0 + 10 \text{ minutes}) =?$

▷ A first guess is that your location will be exactly your *current* location

$$l(t_0 + 10 \text{ minutes}) \approx l(t_0).$$

Given how large the universe is and thus how many possible locations there are, this is certainly not too bad.

▷ If you want to do a bit better than that, you can also include your *current* velocity[1] $\dot{l}(t_0) \equiv \partial_t l(t)\big|_{t_0}$. The total distance you will travel in 10 minutes if you continue to move at your current velocity is this velocity times 10 minutes: $\dot{l}(t_0) \times 10$ minutes . Therefore, your total second estimate is now your current location plus the velocity you are traveling times

[1] Here ∂_t is a shorthand notation for $\frac{\partial}{\partial t}$ and $\partial_t l(t)$ yields the velocity (rate of change). After taking the derivative, we evaluate the velocity function $\dot{l}(t) \equiv \partial_t l(t)$ at t_0: $\dot{l}(t_0) = \partial_t l(t)\big|_{t_0}$.

10 minutes

$$l(t_0 + 10 \text{ minutes}) \approx l(t_0) + \dot{l}(t_0) \times 10 \text{ minutes}. \quad (A.1)$$

▷ If you want to get an even better estimate you need to take into account that your velocity possibly changes. The rate of change of the velocity $\ddot{l}(t_0) = \partial_t^2 l(t)\big|_{t_0}$ is what we call acceleration. So in this third step you take additionally your *current* acceleration into account[2]

$$l(t_0 + 10 \text{ minutes}) \approx l(t_0) + \dot{l}(t_0) \times 10 \text{ minutes} \\ + \frac{1}{2}\ddot{l}(t_0) \times (10 \text{ minutes})^2.$$

▷ Our estimate will still not yield the perfect final location since additionally we need to take into account that our acceleration could change during the 10 minutes. We could therefore take additionally the current rate of change of our acceleration into account.

This game never ends and the only limiting factor is how exact we want to estimate our future location. For many real-world purposes our first order approximation (Eq. (A.1)) would already be perfectly sufficient.

The procedure described above is exactly the motivation behind the Taylor expansion. In general, we want to estimate the value of some function $f(x)$ at some value of x by using our knowledge of the function value at some fixed point a. The **Taylor series** then reads[3]

$$\begin{aligned} f(x) &= \sum_{n=0}^{\infty} \frac{f^{(n)}(a)(x-a)^n}{n!} \\ &= \frac{f^{(0)}(a)(x-a)^0}{0!} + \frac{f^{(1)}(a)(x-a)^1}{1!} + \frac{f^{(2)}(a)(x-a)^2}{2!} \\ &\quad + \frac{f^{(3)}(a)(x-a)^3}{3!} + \dots, \end{aligned} \quad (A.2)$$

This equation is completely analogous to our estimation of the future location above. Here $f(a)$ is the value of the function

---

[2] The factor $\frac{1}{2}$ and that we need to square the 10 minutes follows since to get from an acceleration to a location, we have to integrate twice:

$$\int dt \int dt \ddot{x}(t_0) = \int dt \ddot{x}(t_0) t \\ = \frac{1}{2}\ddot{x}(t_0) t^2$$

where $\ddot{x}(t_0)$ is the value of the acceleration at $t = t_0$ (= a constant).

[3] Here the superscript n denotes the n-th derivative. For example $f^{(0)} = f$ and $f^{(1)}$ is $\partial_x f$.

at the point a we are expanding around. Moreover, $x - a$ is analogous to the timespan 10 minutes above: if we want to know the location at $x = 5\!:\!10$ pm by using our knowledge at $a = 5\!:\!00$ pm we get here $x - a = 5\!:\!10$ pm $- 5\!:\!00$ pm $= 10$ minutes.

To understand the Taylor expansion a bit better it is helpful to look at concrete examples.

We start with one the simplest but most important example: the exponential function. Putting $f(x) = e^x$ into Eq. (A.2) yields

$$e^x = \sum_{n=0}^{\infty} \frac{(e^0)^{(n)}(x-0)^n}{n!}$$

The crucial puzzle pieces that we need are therefore $(e^x)' = e^x$ and $e^0 = 1$. Putting this into the general formula (Eq. A.2) yields

$$e^x = \sum_{n=0}^{\infty} \frac{e^0(x-0)^n}{n!} = \sum_{n=0}^{\infty} \frac{x^n}{n!} \quad (A.3)$$

This result can be used as a definition of e^x.

Next let's assume the function we want to approximate is $\sin(x)$ and we want to expand it around $x = 0$. Putting $f(x) = \sin(x)$ into Eq. (A.2) yields

$$\sin(x) = \sum_{n=0}^{\infty} \frac{\sin^{(n)}(0)(x-0)^n}{n!}$$

The crucial information we therefore need is $(\sin(x))' = \cos(x)$, $(\cos(x))' = -\sin(x)$, $\cos(0) = 1$ and $\sin(0) = 0$. Because of $\sin(0) = 0$ every term with even n vanishes, which we can use if we split the sum. Observe that

$$\sum_{n=0}^{\infty} n = \sum_{n=0}^{\infty} (2n+1) + \sum_{n=0}^{\infty} (2n)$$

$$1+2+3+4+5+6\ldots = 1+3+5+\ldots \quad +2+4+6+\ldots \quad (A.4)$$

Therefore, splitting the sum in terms of even and uneven terms yields

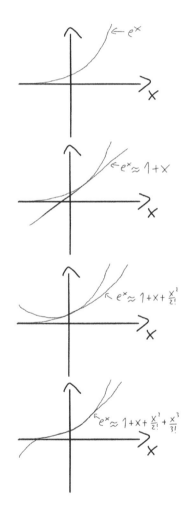

$$\sin(x) = \sum_{n=0}^{\infty} \frac{\sin^{(2n+1)}(0)(x-0)^{2n+1}}{(2n+1)!} + \underbrace{\sum_{n=0}^{\infty} \frac{\sin^{(2n)}(0)(x-0)^{2n}}{(2n)!}}_{=0}$$

$$= \sum_{n=0}^{\infty} \frac{\sin^{(2n+1)}(0)(x-0)^{2n+1}}{(2n+1)!}. \tag{A.5}$$

Moreover, every even derivative of $\sin(x)$, i.e., $\sin^{(2n)}$ is again $\sin(x)$ (with possibly a minus sign in front of it). Therefore the second term here vanishes since $\sin(0) = 0$. The remaining terms are uneven derivatives of $\sin(x)$, which are all proportional to $\cos(x)$. We now use

$$\sin(x)^{(1)} = \cos(x)$$
$$\sin(x)^{(2)} = \cos'(x) = -\sin(x)$$
$$\sin(x)^{(3)} = -\sin'(x) = -\cos(x)$$
$$\sin(x)^{(4)} = -\cos'(x) = \sin(x)$$
$$\sin(x)^{(5)} = \sin'(x) = \cos(x)$$

The general pattern is therefore $\sin^{(2n+1)}(x) = (-1)^n \cos(x)$, as you can check by putting some integer values for n into the formula[4].

[4] $\sin^{(1)}(x) = \sin^{(2\cdot 0+1)}(x) = (-1)^0 \cos(x) = \cos(x)$, $\sin^{(3)}(x) = \sin^{(2\cdot 1+1)}(x) = (-1)^1 \cos(x) = -\cos(x)$

Thus, we can rewrite Eq. A.5 as

$$\sin(x) = \sum_{n=0}^{\infty} \frac{\sin^{(2n+1)}(0)(x-0)^{2n+1}}{(2n+1)!}$$
$$= \sum_{n=0}^{\infty} \frac{(-1)^n \cos(0)(x-0)^{2n+1}}{(2n+1)!}$$
$$\underbrace{=}_{\cos(0)=1} \sum_{n=0}^{\infty} \frac{(-1)^n (x)^{2n+1}}{(2n+1)!} \tag{A.6}$$

This is the Taylor expansion of $\sin(x)$, which we can also use as a definition of $\sin(x)$.

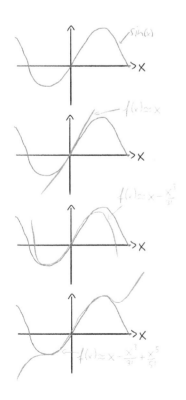

B
Fourier Transform

"One can Fourier transform anything - often meaningfully." - John Tukey

What the Fourier transform does to a function is basically the same that a prism does to sunlight:

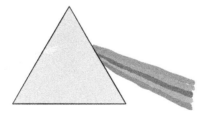

When white sunlight enters a prism we get a rainbow full of colored light[1]. In this sense the white sunlight consists of all this colored light and what the prism does is to reveal these basic constituents.

Now it's not sunlight that enters the Fourier transform, but a function. However, the result is basically the same:

[1] This is a result of the fact that light of different wavelengths get diffracted differently.

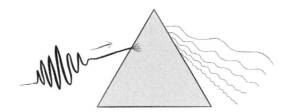

Analogous to the prism here the Fourier transform breaks the incoming function into its basic constituents. The pieces that we get as the result of a Fourier transform are periodic functions (cosines, sines) with different frequencies. The Fourier transform tells us how much of each basic building block is needed to build the original function.

Formulated a bit differently, the basic idea of the Fourier transform is analogous to the idea that we can express any vector $\vec{v}$ in terms of basis vectors ($\vec{e}_1, \vec{e}_2, \vec{e}_3$). The most common choice for these basis vectors are

$$\vec{e}_1 = \begin{pmatrix} 1 \\ 0 \\ 0 \end{pmatrix}, \quad \vec{e}_2 = \begin{pmatrix} 0 \\ 1 \\ 0 \end{pmatrix}, \quad \vec{e}_3 = \begin{pmatrix} 0 \\ 0 \\ 1 \end{pmatrix}. \tag{B.1}$$

We can write any three-dimensional vector $\vec{v}$ in terms of these basis vectors:

$$\vec{v} = \begin{pmatrix} v_1 \\ v_2 \\ v_3 \end{pmatrix} = v_1 \vec{e}_1 + v_2 \vec{e}_2 + v_3 \vec{e}_3$$

$$= v_1 \begin{pmatrix} 1 \\ 0 \\ 0 \end{pmatrix} + v_2 \begin{pmatrix} 0 \\ 1 \\ 0 \end{pmatrix} + v_3 \begin{pmatrix} 0 \\ 0 \\ 1 \end{pmatrix}$$

Now we do the same thing for **real functions**. Here we have infinitely many basis functions: $\sin(kx)$ and $\cos(kx)$, where k is any real number. Using this basis we can write every periodic function $f(x)$ as

$$f(x) = \sum_{k=0}^{\infty} (a_k \cos(kx) + b_k \sin(kx)) \tag{B.2}$$

with constant coefficients a_k and b_k.

For **complex functions** we use the basis e^{ikx} and e^{-ikx} and use an integral instead of a sum[2].

$$f(x) = \int_0^\infty dk \left(a_k e^{ikx} + b_k e^{-ikx} \right), \tag{B.3}$$

which we can rewrite as[3]

$$f(x) = \int_{-\infty}^\infty dk f_k e^{-ikx}. \tag{B.4}$$

Now the expansion coefficients f_k are usually denoted by $\tilde{f}(k)$ and called *the* **Fourier transform** of $f(x)$. It's also possible to invert the whole procedure:[4]

$$\tilde{f}(k) = \frac{1}{2\pi} \int_{-\infty}^\infty dx f(x) e^{ikx}. \tag{B.5}$$

In physical terms we can understand the Fourier expansion as a basis change from the position basis x to the momentum basis k.

[2] Recall that an integral is, in some sense, a finely grained sum.

[3] Take note that now our integral goes from $-\infty$ to ∞.

[4] The factor 2π here is a normalization factor that makes sure that we if combine the Fourier expansion formula and the inverse formula we really up with the function we started with.

C

Delta Distribution

The easiest way to understand the Delta distribution[1] (or Dirac delta) is to recall a simpler but analogous mathematical object: the **Kronecker delta** δ_{ij}, which is defined as follows:

$$\delta_{ij} = \begin{cases} 1 & \text{if } i = j \\ 0 & \text{if } i \neq j \end{cases} \tag{C.1}$$

[1] The Delta distribution is not really a function in the strict mathematical sense and therefore a new word was invented: distributions.

In matrix form, the Kronecker delta is simply the unit matrix[2]. The Kronecker delta δ_{ij} is useful because it allows us to pick one specific term of any sum. For example, let's consider the sum

$$\sum_{i=1}^{3} a_i b_j = a_1 b_j + a_2 b_j + a_3 b_j \tag{C.3}$$

[2] For example, in two-dimensions

$$1_{(2\times 2)} = \begin{pmatrix} 1 & 0 \\ 0 & 1 \end{pmatrix}. \tag{C.2}$$

and let's say we want to extract only the second term. We can do this by multiplying the sum with the Kronecker delta δ_{2i}:

$$\sum_{i=1}^{3} \delta_{2i} a_i b_j = \underbrace{\delta_{21}}_{=0} a_1 b_j + \underbrace{\delta_{22}}_{=1} a_2 b_j + \underbrace{\delta_{23}}_{=0} a_3 b_j = a_2 b_j. \tag{C.4}$$

In general, we have

$$\sum_{i=1}^{3} \delta_{ik} a_i b_j = a_k b_j. \tag{C.5}$$

The **delta distribution** $\delta(x-y)$ is a generalization of this idea for integrals instead of sum. So concretely this means that we can use the delta distribution to extract specific terms from any given integral:

$$\int dx f(x)\delta(x-y) = f(y). \tag{C.6}$$

In words this means that the delta distribution allows us to extract exactly one term - the term $x=y$ - from the infinitely many terms that we means by the integral sign. For example,

$$\int dx f(x)\delta(x-2) = f(2).$$

Now, one example where the Kronecker delta appears is

$$\frac{\partial x_i}{\partial x_j} = \delta_{ij}. \tag{C.7}$$

The derivative of $\partial_x x = 1$, whereas $\partial_x y = 0$ and $\partial_x z = 0$.

Completely analogous the delta distribution appears as follows:

$$\frac{\partial f(x_i)}{\partial f(x_j)} = \delta(x_i - x_j). \tag{C.8}$$

The delta distribution is also often introduced by the following definition

$$\delta(x-y) = \begin{cases} \infty & \text{if } x=y, \\ 0 & \text{if } x \neq y \end{cases}, \tag{C.9}$$

which is somewhat analogous to the definition of the Kronecker delta in Eq. (C.1). Moreover, when we use a constant function in Eq. (C.6), for example, $f(x) = 1$ we get the following remarkable equation

$$\int dx 1 \delta(x-y) = 1. \tag{C.10}$$

The thing is that Eq. (C.6) tells us that if we have the delta distribution $\delta(x-y)$ together with a function under an integral, the result is the value of the function at $y = x$. Here we have a constant function and its value at $y = x$ is simply 1.

In words these properties mean that the delta distribution is infinitely thin (only nonzero at $y = x$) but also infinitely high function that yields exactly one if we integrate it all over space.

This is also why the delta distribution is so important in Quantum Mechanics. Here we describe particles using wave packets and the delta distribution is like an infinitely thin wave packet. In physical terms this means that the delta distribution describes a particle that is localized at exactly one point without any uncertainty. However, at the same time the momentum uncertainty is infinite. This comes about since we can understand the delta distribution as a infinite sum of all possible plane waves[3]. Each plane wave corresponds to exactly one specific momentum value. Since we need all possible plane waves, the delta distribution is a superposition of all possible momentum values:

[3] A plane wave looks mathematically like this: e^{ixk}.

$$\delta(x - y) = \frac{1}{2\pi} \int_{-\infty}^{\infty} e^{i(x-y)k} dk. \quad (C.11)$$

This is known as the **integral representation** of the delta distribution.

This can be motivated by putting the formula for the inverse Fourier transformation (Eq. (B.5))

$$\tilde{f}(k) = \frac{1}{2\pi} \int_{-\infty}^{\infty} dx f(x) e^{ikx}. \quad (C.12)$$

into the normal Fourier expansion formula (Eq. (B.4)):

$$f(x) = \int_{-\infty}^{\infty} dk \tilde{f}(k) e^{-ikx}$$
$$= \int_{-\infty}^{\infty} dk \left(\frac{1}{2\pi} \int_{-\infty}^{\infty} dx' f(x') e^{ikx'} \right) e^{-ikx}$$
$$= \int_{-\infty}^{\infty} dx' f(x') \left(\int_{-\infty}^{\infty} \frac{dk}{2\pi} e^{ik(x'-x)} \right)$$

Now, compare this with the defining equation (Eq. (C.6)) which we recite here for convenience[4]

[4] We changed the notation $x \to x'$ and $y \to x$ to make the analogy clearer.

$$f(x) = \int dx' f(x') \delta(x' - x). \quad (C.13)$$

If we identify the last objects in these equations, we end up with the integral representation of the delta distribution (Eq. (C.11)).

Bibliography

J. E. Baggott. *The quantum story : a history in 40 moments.* Oxford University Press, Oxford England New York, 2011. ISBN 978-0199566846.

L. E. Ballentine. The statistical interpretation of quantum mechanics. *Rev. Mod. Phys.*, 42:358–381, 1970. DOI: 10.1103/RevModPhys.42.358.

Leslie Ballentine. *Quantum mechanics : a modern development.* World Scientific, Singapore River Edge, NJ, 2000. ISBN 9789810241056.

Adam Becker. *What is real? : the unfinished quest for the meaning of quantum physics.* Basic Books, New York, NY, 2018. ISBN 978-0465096053.

Felix Bloch. Heisenberg and the early days of quantum mechanics. *Physics Today*, December 1976.

David Bohm. *Quantum theory.* Dover Publications, New York, 1989. ISBN 978-0486659695.

Siegmund Brandt. *The picture book of quantum mechanics.* Springer, New York, NY, 2012. ISBN 9781461439509.

Bernard d'Espagnat. *Conceptual foundations of quantum mechanics.* Advanced Book Program, Perseus Books, Reading, Mass, 1999. ISBN 978-0738201047.

P. A. M. Dirac. *The principles of quantum mechanics.* Clarendon Press, Oxford England, 1981. ISBN 9780198520115.

Richard Feynman. *Quantum mechanics and path integrals.* Dover Publications, Mineola, N.Y, 2010. ISBN 978-0486477220.

Richard Feynman. *The Feynman lectures on physics.* Basic Books, a member of the Perseus Books Group, New York, 2011. ISBN 978-0465025015.

Richard Feynman. *QED : the strange theory of light and matter.* Princeton University Press, Princeton, NJ, 2014. ISBN 978-0691164090.

Christopher A. Fuchs and Asher Peres. Quantum theory needs no 'interpretation'. *Physics Today*, 53(3):70–71, March 2000. DOI: 10.1063/1.883004.

V. M. Galitski. *Exploring quantum mechanics : a collection of 700+ solved problems for students, lecturers, and researchers.* Oxford University Press, Oxford, 2013. ISBN 9780199232727.

F. Gieres. Dirac's formalism and mathematical surprises in quantum mechanics. *Rept. Prog. Phys.*, 63:1893, 2000. DOI: 10.1088/0034-4885/63/12/201.

David Griffiths. *Introduction to quantum mechanics.* Pearson Prentice Hall, Upper Saddle River, NJ, 2005. ISBN 9780131118928.

Brian Hall. *Quantum theory for mathematicians.* Springer, New York, 2013. ISBN 9781489993625.

Peter Holland. *The quantum theory of motion : an account of the de Broglie-Bohm causal interpretation of quantum mechanics.* Cambridge University Press, Cambridge England New York, NY, 1995. ISBN 978-0521485432.

E. T. Jaynes. *Probability Theory as Logic*, pages 1–16. Springer Netherlands, Dordrecht, 1990. ISBN 978-94-009-0683-9.

Robert Klauber. *Student friendly quantum field theory : basic principles and quantum electrodynamics.* Sandtrove Press, Fairfield, Iowa, 2014. ISBN 9780984513956.

Sanjoy Mahajan. *The art of insight in science and engineering : mastering complexity.* The MIT Press, Cambridge, Massachusetts, 2014. ISBN 978-0262526548.

Albert Messiah. *Quantum mechanics*. North-Holland, Amsterdam, 1965. ISBN 9780471597667.

J. E. Moyal. Quantum mechanics as a statistical theory. *Proc. Cambridge Phil. Soc.*, 45:99–124, 1949. DOI: 10.1017/S0305004100000487.

Edward Nelson. Review of stochastic mechanics. *Journal of Physics: Conference Series*, 361(1):012011, 2012. URL http://stacks.iop.org/1742-6596/361/i=1/a=012011.

Hrvoje Nikolic. Quantum mechanics: Myths and facts. *Found. Phys.*, 37:1563–1611, 2007. DOI: 10.1007/s10701-007-9176-y.

Travis Norsen. *Foundations of quantum mechanics : an exploration of the physical meaning of quantum theory*. Springer, Cham, Switzerland, 2017. ISBN 978-3319658667.

Roland Omnes. *The interpretation of quantum mechanics*. Princeton University Press, Princeton, N.J, 1994. ISBN 978-0691036694.

Yoav Peleg. *Quantum mechanics : based on Schaum's outline of theory and problems of quantum mechanics*. McGraw-Hill, New York, 2006. ISBN 9780071455336.

Roger Penrose. *The emperor's new mind : concerning computers, minds and the laws of physics*. Oxford University Press, Oxford, 2016. ISBN 9780198784920.

Asher Peres. *Quantum theory : concepts and methods*. Kluwer Academic, Dordrecht Boston, 1993. ISBN 978-0-7923-2549-9.

Carlo Rovelli. Space is blue and birds fly through it. 2018. DOI: 10.1098/rsta.2017.0312.

J. J. Sakurai. *Modern quantum mechanics*. Pearson India Education Services, Noida, India, 2014. ISBN 9789332519008.

Paul Schilpp. *Albert Einstein, philosopher-scientist*. Open Court, La Salle, Ill, 1970. ISBN 979-0875482865.

Maximilian Schlosshauer. *Elegance and enigma : the quantum interviews*. Springer, New York, 2011. ISBN 978-3642208799.

Jakob Schwichtenberg. *Physics from Symmetry.* Springer, Cham, Switzerland, 2018. ISBN 978-3319666303.

James Sethna. *Statistical mechanics : entropy, order parameters, and complexity.* Oxford University Press, Oxford New York, 2006. ISBN 9780198566779.

Ravi Shankar. *Principles of Quantum Mechanics.* Springer Verlag, City, 2014. ISBN 9781461576754.

F Strocchi. *An introduction to the mathematical structure of quantum mechanics : a short course for mathematicians.* World Scientific, Hackensack, N.J, 2008. ISBN 9789812835222.

Claude Tannoudji. *Quantum mechanics.* Wiley, New York, 1977. ISBN 9780471164333.

Bernd Thaller. *Visual quantum mechanics : selected topics with computer-generated animations of quantum-mechanical phenomena.* Springer/TELOS, New York, 2000. ISBN 9780387989297.

Peter Woit. *Quantum theory, groups and representations : an introduction.* Springer, Cham, Switzerland, 2017. ISBN 9783319646107.

Cosmas Zachos. *Quantum mechanics in phase space : an overview with selected papers.* World Scientific, New Jersey London, 2005. ISBN 9812383840.

A Zee. *Quantum field theory in a nutshell.* Princeton University Press, Princeton, N.J, 2010. ISBN 9780691140346.

Index

Angular Momentum, 83
    Algebra, 84
        Commutation Relation, 84
        Operator, 84
Angular Momentum Algebra, 83

Basis Change, 62
Basis Expansion, 57
Born Approximation, 172

Canonical Commutation Relation, 75
Classical Limit, 93, 209
Classical Path, 209
Clebsch-Gordan Coefficients, 156
Commutator, 74
Configuration Space, 182
Constructive Interference, 211
Continuity Equation, 217

Destructive Interference, 211
Double Slit Experiment
    Path Integral Description, 199
    Pilot Wave Description, 196
Double-Slit Experiment, 33

Ehrenfest's theorem, 94
Eigenvalue, 82
Eigenvector, 82
Energy Operator, 71, 72
Euler's Formula, 206
Everyday Space, 181

Expectation Value, 51
Expectation Value Quantum Mechanics, 58

Fermi's Golden Rule, 172
Fourier Transform, 249

Gaussian Wave Packet, 81
Generator, 71
Group Theory, 68

Hamilton Operator, 77
Hamiltonian Formulation, 190
Harmonic Oscillator, 137
    Algebraic Method, 140
    Driven, 161
Heisenberg Mechanics, 223
Hilbert Space, 189
Hydrogen Atom, 124

Interpretations of Quantum Mechanics, 227
    Copenhagen, 229
    Many Worlds, 230
    Statistical, 232
    Stochastic, 232

Ket, 47
Koopman-von-Neumann Formulation, 190

Ladder Operators, 140

Lagrangian Formulation, 190
Liouville's Equation, 216

Mathematical Arenas, 181
Momentum Operator, 71, 72
Moyal Bracket, 222

Newtonian Formulation, 190
Noether's Theorem, 66

Partial Wave Analysis, 172
Particle
  Box Scattering, 127
  Finite Box, 118
  Free, 79
  Infinite Box, 113
    Perturbations, 170
  Spherical Potential, 158
  Three Dimensional Box, 158
Path Integral, 203
Path Integral Formulation, 190, 198, 225
Pauli Matrices, 85
Perturbation Theory, 164
  Degenerate, 171
  General, 165
  Time-Dependent, 171
Phase Space, 187
  Filamentation, 220
  Flow, 214
    Compressible, 219
    Incompressible, 218
Phase Space Formulation, 190, 213, 225
Pilot-Wave Formulation, 190, 193, 225
Planck Constant, 73
Plank Constant, 72
Poisson Bracket, 215
  Origin, 222

Probability Density, 215
  Explicit, 217
Product Space, 183

Quantum Numbers, 90
Quantum Operators, 66
Quantum Pendulum, 160
Quantum Potential, 196
Quantum Tunneling, 120
Quantum Waves, 79

Scattering Amplitude, 172
Schrödinger Equation
  Classification of Solutions, 109
  Stationary, 107
  Time-Independent, 107
Schrödinger equation, 77
Schrödinger's Cat, 228
Spin, 85, 147
  Addition, 154
  Algebra, 89
  Commutation Relation, 89
  Measurements, 149
State Vector, 47
States, 47
Statistical Tools, 54
Stern-Gerlach Experiment, 88, 149
Superposition, 80
Symmetries, 67

Taylor Expansion, 245
Time Evolution, 75
Time Evolution Operator, 78

Uncertainty, 43

Wave Function, 60
Wave Function Formulation, 190, 225
WKB Method, 175

Made in the USA
Middletown, DE
27 March 2019